MATHEMATICS FOR THE MARINER

DENNIS A. J. MOREY

AND

THEODORE W. THROCKMORTON

authorHOUSE™

1663 LIBERTY DRIVE, SUITE 200
BLOOMINGTON, INDIANA 47403
(800) 839-8640
WWW.AUTHORHOUSE.COM

First published by AuthorHouse 11/01/04

ISBN: 1-4208-0939-3 (sc)

Printed in the United States of America
Bloomington, Indiana

This book is printed on acid-free paper.

Contents.

𝔐𝔞𝔱𝔥𝔢𝔪𝔞𝔱𝔦𝔠𝔰 𝔣𝔬𝔯 𝔱𝔥𝔢 𝔐𝔞𝔯𝔦𝔫𝔢𝔯.

Introduction:

This presentation is made to satisfy the curiosity of those amateur boaters who wish a better understanding of the navigational techniques employed by the early adventurers and circumnavigators in the days prior to the modern, sophisticated instruments, and to appreciate the contributions made by the great explorers, such as Drake, Cook, Bligh and more recently, Chichester.

It is hoped that this will also demonstrate that essentially all that is accomplished in chart plotting may be achieved mathematically, with a greater degree of accuracy, and with the modern calculator much faster. Nevertheless, it would be foolhardy to rely on these calculations only. A plot should always be run, using the calculations as confirmation.

A third purpose is an attempt to encourage amateur navigators to take pride in this erudition, and offset the current downward trend in educational standards, not only within our ranks, but nationwide in all fields.   This tendency has been further promoted by the introduction of inexpensive Loran and GPS, which has led to an irrational dependence upon these electronic gadgets to the exclusion of charting and other modes of navigation.

Common sense dictates that we should not rely exclusively on this modern technology, but be prepared, in the event the "fuse blows" to employ the time-tested methods of the past.

Before embarking upon the exploration of the mathematics of navigation, the reader is urged to obtain a low-cost, simple scientific calculator.   This instrument will replace the need to spend a great deal of time extracting figures from tables.

The instrument recommended for this course is the TI-30x, either battery operated or solar powered. However, other similar instruments are available, but be sure that they have the trigonometric keys, the reciprocal key (1/x), the conversion for degrees, minutes and seconds to decimals (DMS-DD) and vice versa, the rectangular to polar (R-P) and vice versa, as well as logs, square and square root.   Spend a little time familiarizing yourself with the functions of the calculator, so that they will not be a distraction as you tackle the following navigational problems.

Before we embark upon the substance of this course, we will clarify certain procedures employed, to avoid distraction and confusion.

1.   The Arrow Convention:

This is a technique to depict the various lines on the plotted chart, and is used by most air and marine navigators throughout the world.

(a)   The True Course being steered by the helmsman is shown with a single arrow pointing the direction of the heading.

—————————⟶—————————

(b)   The Track Made Good is depicted with 2 arrows in the direction of motion.

—————————⟫—————————

(c)   The Set and Drift vector is depicted with 3 arrows, pointing in the direction of current flow.

Thus the navigational triangle would be "labeled" as seen below:

This technique makes each line clearly evident and avoids cluttering the chart with data which is always contained in the log.

2.   Interior Angles within the Navigational Triangle:

The mathematical solutions of the navigational triangle determine the lengths of the sides and the angles subtended within the triangle by those sides. These angles are related to the course and track, but are not necessarily equal to those headings.

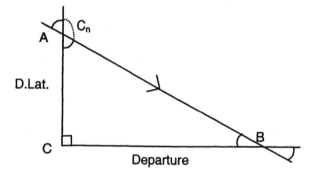

The course angle at A (CAB) is computed and, in this case, must be subtracted from 180° to obtain the course to steer, i.e., C_n.

If the interior angle at B is determined, then the C_n is equal to 90° + B.

In the case of the spherical triangle, the same principle holds true, except that the reference lines are the meridians from the elevated pole through the origin and destination, respectively.

If the interior angle C is computed,

then $C_n = 360° - C$.

If B is Computed,

then $C_n = 180° + B$

Note: All meridians run North-South, i.e., 000°-180°

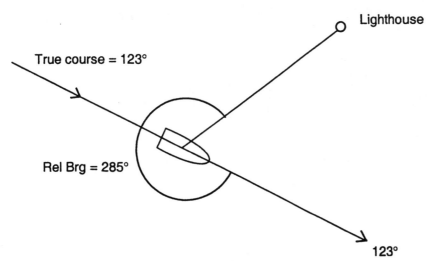

3. Bearings

These measurements must always be converted to true bearings before plotting, i.e., from compass, magnetic or relative.

Except for astronomical azimuths, all bearings are made upon easily identifiable landmarks or buoys, which may then be converted to the reciprocal in order to be plotted from the landmark, etc., as a position line.

A plea is made here for the more frequent use of relative bearings, which can be made with the pelorus or "dumb compass." This instrument is inexpensive, durable, has no intrinsic errors, and is simple to use. The only requirement is to align the 0°-180° axis parallel to the vessel's keel. Because of visual obstruction due to the superstructure, it may be necessary to set up 2 of these instruments to obtain bearings in any direction.

The relative bearing is measured from the bow, clockwise through 360°. Thus a stern bearing would be 180°, while the beam bearings would be 090° to starboard, and 270° to port.

The relative bearing is added to the true course steered to obtain the true bearing of the object, and its reciprocal obtained to plot from the object on the chart.

Example:

A lighthouse bears 285° relative. Your course is 123° T. What is the plotted bearing from the lighthouse?

Relative bearings	=	285° R
True course	=	123° T
True bearing	=	408° – 360° = 048° T
Reciprocal	=	048° + 180° = 228° T

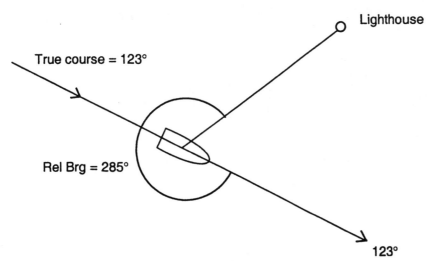

Example:

You are on a course of 333° T. A prominent landmark bears 076°
Relative. What is the reciprocal true bearings to be plotted from the landmark?

Course   =   333° T
Bearing   =   076° R
True bearings   =   409° – 360° = 049° T

Reciprocal   =   049° + 180° = 229° T

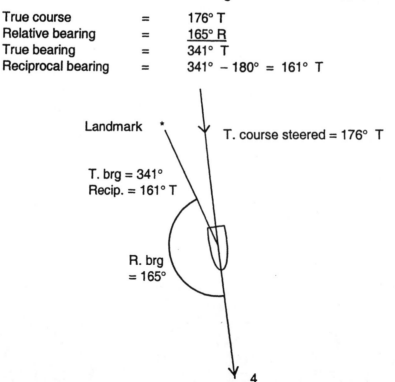

333° T

076° R

* Landmark

T. brg. = 049° T

Recip. = 229° T

True course = 333° T

Example:

Your course steered is 176° T. Relative Bearing of a landmark is 165° R.

True course   =   176° T
Relative bearing   =   165° R
True bearing   =   341° T
Reciprocal bearing =   341° – 180° = 161° T

Landmark   *

T. course steered = 176° T

T. brg = 341°
Recip. = 161° T

R. brg
= 165°

4

The pelorus also enables you to take back bearings of objects close to which you passed earlier, and thus gives you the angle of drift, i.e., the drift between the course steered and the track made good.

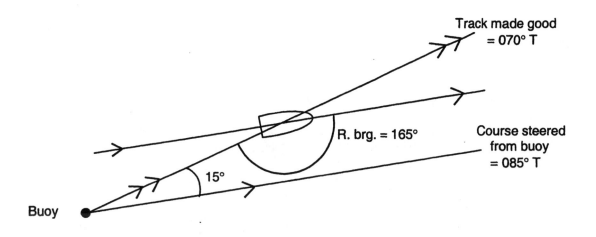

Converting a Course Angle to a True Bearing:

Example:

You are on a course of 083° T. You calculate a course angle between the bearing and your course of 47°. What is the bearing you would plot from the landmark observed?

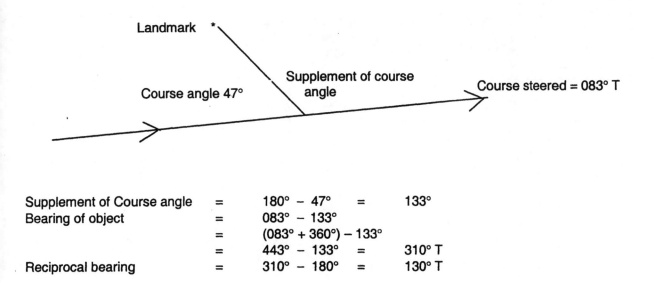

Supplement of Course angle	=	180° − 47°	=	133°
Bearing of object	=	083° − 133°		
	=	(083° + 360°) − 133°		
	=	443° − 133°	=	310° T
Reciprocal bearing	=	310° − 180°	=	130° T

5

Chapter 1.

Section 1

This addresses the subject of plane trigonometry, and includes all those ratios and formulae which have a direct application to the solution of navigational problems.

This section is restricted to the solution of plane triangles, i.e. where the earth's surface is considered flat. Such applications are limited to distances less than 600 nautical miles.

Basic Trigonometry:

This is a system in which angles are expressed as ratios of two sides of a right triangle, i.e. a triangle in which one angle is a right angle (90 degrees). Each angle is lettered with a capital, and the side opposite this angle is designated with its small letter.

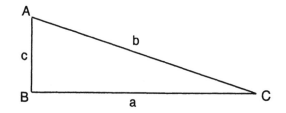

Trig. Functions:

$$
\begin{aligned}
\text{Sine of angle C} &= \text{opposite / hypotenuse} &= c / b \\
\text{Cosine of C} &= \text{adjacent / hypotenuse} &= a / b \\
\text{Tangent of C} &= \text{opposite / adjacent} &= c / a
\end{aligned}
$$

The reciprocals of these functions are:

$$
\begin{aligned}
\text{Cosecant of C} &= & \text{hypotenuse / opposite} &= b / c \\
\text{Secant of C} &= & \text{hypotenuse / adjacent} &= b / a \\
\text{Cotangent of C} &= & \text{adjacent / opposite} &= a / c
\end{aligned}
$$

*Note:

Since your calculator does not have these reciprocal functions, simply use the reciprocal button (1/x) to obtain the values of the reciprocals.

The 3 angles of a plane triangle always add up to 180 degrees.

$$A + B + C = 180°$$

Since $B = 90°$, then $A + C = 90°$ (i.e. they are said to be complementary.)

If a = 4, and c = 5. The Tan C = 5 / 4 = 1.25, and C = 51.34°

Then A = 90° - 51.34° = 38.66°
Sin A = 4 / b or b = 4 / sin A = 4 / sin 38.66° = 6.40
Cos 5l.34° = 0.62469 = 4 / 6.4 = 0.625 (rounded off).

*Note:

The sine of an angle = the cosine of its complement.
e.g: sine 30° = cosine 60°
(confirm this with your calculator).

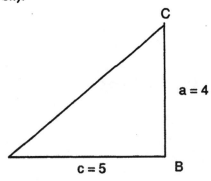

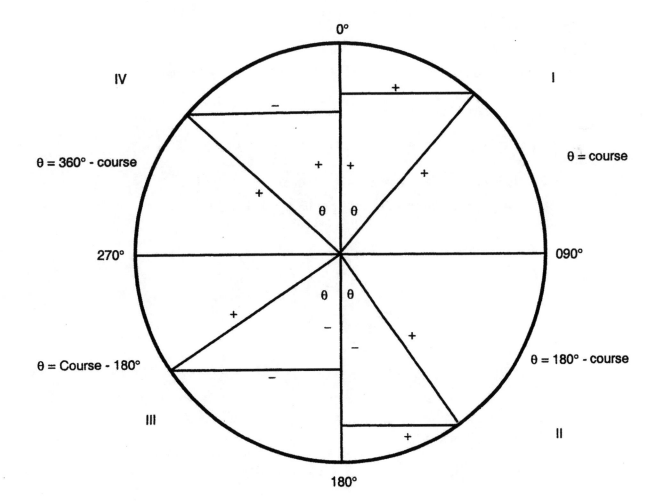

In navigation we use angles from 000° to 360°, i.e. through the 4 quadrants.   Their trig. ratios are identified by changing signs (+ or −).

In the above diagram all radii are considered +.   All lines vertically above the horizontal axis (90° - 270°) are +, and those below -.   All lines horizontal to the vertical axis (0° - 180°) are + if to the right, and - if to the left.

Positive/negative diagram:

	I	II	III	IV	
Sine	+	+	−	−	Cosecant
Cosine	+	−	−	+	Secant
Tangent	+	−	+	−−	Cotangent

Negative values:

$$\sin(-\theta) = -\sin\theta$$
$$\cos(-\theta) = \cos\theta$$
$$\tan(-\theta) = -\tan\theta \quad (2^{nd}\ quadrant)$$
$$\tan(-\theta) = \tan(360° - \theta)(4^{th}\ quadrant)$$

*Useful trig. formulas:

(a) The Sine Law:

(For any oblique triangle (plane) , i.e. one in which there is no right angle):

$$\frac{a}{\sin A} = \frac{b}{\sin B} = \frac{c}{\sin C}$$

*Example:

If B = 25° , b = 4, C = 42° , what is the length of c?

$$\frac{4}{\sin 25°} = \frac{c}{\sin 42°} \quad \text{Therefore } c = \frac{4 \times \sin 42°}{\sin 25°} = 6.3332$$

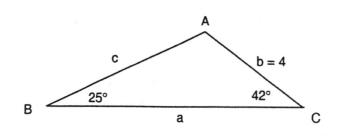

(b) The Law of Cosines:

$$a^2 = b^2 + c^2 - 2 \times b \times c \times \cos A$$

*Example:

Let A = 82°, b = 6, c = 4.   What is a?

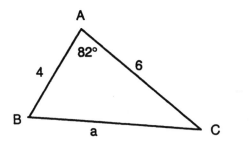

$$a^2 = 6^2 + 4^2 - 2 \times 6 \times 4 \; Cos \; 82°$$
$$= 36 + 16 - 48 \times 0.139173$$
$$= 45.319691$$
$$a = 6.73199$$

Note:

> This Law of Cosines may also be transformed to:

$$Cos \; A = \frac{c^2 + b^2 - a^2}{2 \times b \times c}$$

to solve a plane triangle with only the three sides known.

**Other useful formulas of oblique plane triangles may be found in Table 142a, page 410 , Bowditch , vol. II, 1975 edition.

Trig. Identities:

In addition to the Sine and Cosine laws, there are other trig. identities which are useful in the solution of navigational problems involving plane, oblique triangles.   The identity most frequently used is as follows:

$$Tan \; A = \frac{a \times Sin \; C}{b - a \times Cos \; C}$$

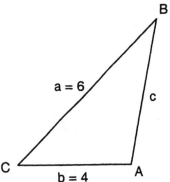

Example:

If C = 40°, a = 6 units, and b = 4 units.   Solve for A.

$$Tan \; A = \frac{6 \times Sin \; 40°}{4 - 6 \times Cos \; 40°} = \frac{3.85673}{-0.59627} = -6.46813$$

$$A = -81.2° = 98.8°$$

*Note:

The presence of a minus here indicates that the angle A is in the 2nd quadrant, i.e. between 90° and 180°.   In other words, it is the supplement:

$$180° - 81.2° = 98.8°$$

*Remember:

Sines in the 2nd quadrant are positive, and therefore a source of ambiguity.   This is not the case with Cosines, where 2nd quadrant values are negative, and the calculator can differentiate.

2nd quadrant tangents are equal numerically to their 1st quadrant supplements, but the calculator can only differentiate by the minus sign, and you must do the rest.

Section 2:

This section contains practical applications to a variety of navigational problems.

The Application of Plane Trigonometry to Navigational Problems:

The difference between two adjacent meridians of longitude is measured in degrees and minutes of arc. This remains unchanged no matter what the latitude might be.   However, the distance between these two meridians in nautical miles varies from a maximum at the Equator to zero at the poles, due to the convergence of these meridians towards either pole.   One minute of DLo. at the Equator = 1 nautical mile, diminishing to zero as one progresses towards the poles.

At the Equator 1° of DLo. = 60 n .m.
At 30° N. this same arc = 51.96 n.m.
At 60° N. = 30 n.m.   At 75° N. = 15.53 n.m.

Most courses do not stay on the same parallel of latitude, but cross the meridians obliquely. To allow for this change of latitude in the conversion from DLo. to departure (n.m's) an average or mid-latitude (Lm) is calculated..

The conversion of DL0. to p (departure) and vice-versa is easily accomplished with the following right triangle in which the known angle is Lm., the hypotenuse = the DLo. and the adjacent = p.

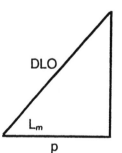

Thus:

Cos Lm = p / DLo.

or   p = DLo. x Cos Lm
or   DLo = p / Cos Lm = p x Sec Lm

Example:

You sail from 35° N., 75° W. to 41° N., 67° W.

What is your departure ?

D. Lat. = 6°   Therefore the Lm = 38°
D. Long. = 8° = 480′ of arc

Example:

You leave 40° N, 60° W and sail 60 n.m. on a course of 042° T..
What is the longitude at the end of the run?

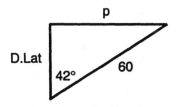

D. lat = 60 x Cos 42° = 44.58869′
p = 60 x Sin 42° = 40.14784 n.m.
L_m = 40° + (0.5 x 44.58869′) = 40.37157°

Therefore DLo. = 40.14784 / Cos 40.37157° = 52.69715′ of arc.

Then Longitude = 60° W − 52.7′ = 59° 07.3′ W

To determine Track and Ground Speed:

You wish to make good a track from A to C of 070° T. Your indicated speed = 8 knots. If the set and drift is 180° T / 2 knots, what true course must you steer to make good the desired track, and what will be your ground speed ?

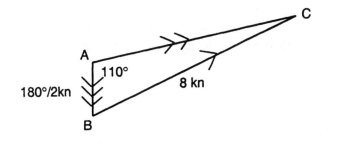

Sin C / 2 = Sin 110° / 8
Then C = 13.58709° = 13.6° (rounded off)
    Therefore Co. to steer = 070° − 13.6° = 056.4° T.
    A + C = 110° + 13.6° = 123.6°
    Then B = 180° − 123.6° = 56.4°

11

To determine the ground speed:

$$b / \text{Sin } 56.4° = 8 / \text{Sin } 110°$$
$$b = (8 \times \text{Sin } 56.4°) / \text{Sin } 110°$$
$$\text{Therefore } b = 7.09101 = 7.1 \text{ knots.}$$

To determine Set and Drift:

You are sailing a course = 100° T., at an indicated speed of 8 knots.   A fix shows that your track made good = 093° T., and that you are making good a ground speed of 7.6 knots.   What is the set and drift?

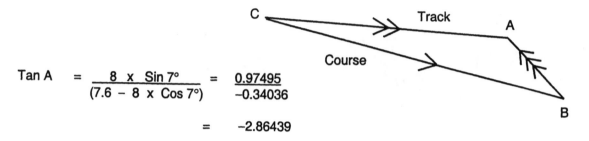

$$\text{Tan } A = \frac{8 \times \text{Sin } 7°}{(7.6 - 8 \times \text{Cos } 7°)} = \frac{0.97495}{-0.34036}$$

$$= -2.86439$$

Therefore A = -70.75516°   The minus indicates a second quadrant angle, i.e. 180° - 70.75516° = 109.24484°.

Then the set = 343.75516° = 343.8° T.

Drift:

$$\frac{c}{\text{Sin } 7°} = \frac{8}{\text{Sin } 109.24484°}$$

$$c = 1.03266$$

Therefore the drift = 1.0 knots.
Set / Drift = 343.8° T / 1.0 knots.

Running Fix :(solved by plane trig)

You are sailing on a course of 087° T. at 8 knots.   At 1000 LST a lighthouse bears 028° T.   At 1015 LST the same lighthouse bears 323° T.   Determine your fix as a bearing and distance from the lighthouse.

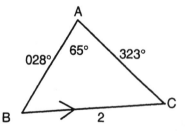

Angle B = 087° - 028° = 59°   Angle C = 323° - (180° + 087°) = 56°
BC = distance run in 15 minutes = 2 n.m.

By Sine Law:

$$2 / \text{Sine } 65° = b / \text{Sin } 59°$$
$$b = (2 \times \text{Sin } 59°) / \text{Sine } 65° = 1.89156 \text{ n.m.}$$

Therefore the fix is: bears 143° T from lighthouse, 1.89 n.m.

To determine the Set and Drift by back bearing on 1st. leg, and a true bearing at end of 2nd. leg:

*Example:

Set course on 100° T., from a given buoy, A, at 8 knots and run for 6 minutes, during which time you obtain a relative back bearing of the buoy by pelorus, thus establishing your drift, e.g. 10° to port. At 6 minutes turn onto the second leg with a course of 190° T, and run for an additional 6 minutes. At the end of this 2nd leg obtain a true bearing of the buoy.

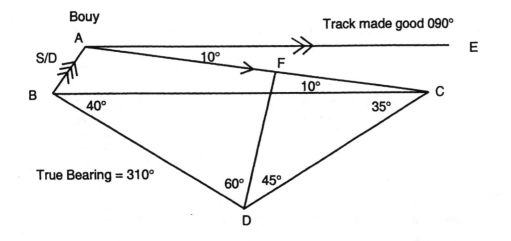

Back bearing on 1st leg = 170° rel. i.e. 10° of port drift.
BC is drawn parallel to AE, i.e. angle ACB = 10°
AF, the 1st leg, is extended to C, so that FC = AF, and also equals FD.

Since FDC is an isosceles triangle, and since DFC is a right angle, then angle DCF = 45°. Since AE and BC are parallel, angle ACB = 10°, and therefore angle BCD = 45° – 10° = 35°. AF, FC and FD all represent runs of 6 minutes at 8 knots, i.e. 0.8 n.m.

Then since triangle FDC is a right triangle the length of DC may be calculated by Pythagorus and found to be 1.13137 n.m. It could also have been determined by the sine law and simple trig.

Solution:

In triangle BCD
$$BC / \sin 105° = 1.13137 / \text{Sin } 40°$$
$$BC = 1.70013$$

In triangle ABC:

$$\text{Tan BAC} = \frac{1.70013 \times \text{Sin } 10°}{1.6 - 1.70013 \times \text{Cos } 10°}$$

$$= \frac{0.29522}{-0.07430}$$

$$= -3.97335$$

BAC = −75.87337 = 104.12663°

Therefore   BAC = 104.1°, and set = 024.1° T

Drift:

drift / Sin 10° = 1.70013 / Sin 104.1°
   i.e., drift = 0.30440 n.m. in 12 minutes.
          = 1.52198 knots.

Set and Drift by back-bearing, quick turn onto Reciprocal Track (not course) and timed return to the starting point, e.g. a buoy.

Example:

You set course from a buoy on a heading of 095° T for 10 minutes at 10 knots (1.66666n.m.), during which time you take a series of back-bearings of the buoy until you are satisfied that you have a consistent reading. From this back-bearing you determine your drift angle and then make a rapid change of course to the reciprocal of your earlier course plus or minus double the angle of drift. This new course should place you on a reciprocal track to the buoy. You measure the time on the return leg to the buoy.

Let us assume that you obtain a consistent back-bearing of 168° relative.   This will give you a 12° port drift..   Your new course back to the buoy will be 095° + 180° - 24° = 251° T

Let us also assume that the time back to the buoy at 10 knots = 9.5 minutes, i.e. 1.58 n.m.

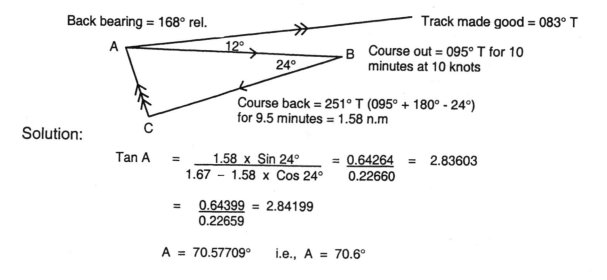

Back bearing = 168° rel.   Track made good = 083° T

A   12°   B   Course out = 095° T for 10
                    24°   minutes at 10 knots

                    Course back = 251° T (095° + 180° - 24°)
                    for 9.5 minutes = 1.58 n.m

C

Solution:

$$\text{Tan A} = \frac{1.58 \times \text{Sin } 24°}{1.67 - 1.58 \times \text{Cos } 24°} = \frac{0.64264}{0.22660} = 2.83603$$

$$= \frac{0.64399}{0.22659} = 2.84199$$

A = 70.57709°   i.e., A = 70.6°

Therefore the set = 345.6° T

Drift / Sin 24° = 1.58 / Sin 70.6° = 0.68132 n.m. in 19.5 minutes.

Therefore the drift = 2.09639 knots.
           Set & Drift = 345.6° T / 2.1 knots.

A Concession to Hi-Tech Nerds:

You are sailing on a course = 118° T at 12 knots. At 1015 your GPS gives your position as 37° 30.9' N., 75° 01.4' W.   30 minutes later your position is 37° 27.4' N., 74° 54.6' W. Determine your track, ground speed and the set and drift.

Solution:

37° 30.9' N	75° 01.4' W
37° 27.4' N	74° 54.6' W
3.5' S	6.8' E

L_m = 37° 29.2' = 37.48667°
p = 6.8 x Cos 37.48667° = 5.39577

        Tan A = 5.39577 / 3.5
             A = 57.03029°

Therefore track = 122.96971° = 123.0°

Ground speed = $\sqrt{3.5^2 + 5.3958^2}$ = 6.43215 n.m. in 30 minutes.
That is, ground speed = 12.86307 knots.

Set & Drift:

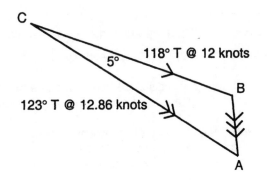

Tan A = 12 x Sin 5° / 12.86 - 12 x Cos 5°
          A = 49.10928° Then B = 125.9°
Therefore set = 172.1° T

Drift:

          AB = 12 x Sin 5° / Sin 49.10928°   = 1.38349
          Therefore Set / Drift = 172.1° T. / 1.4 knots.

The Improved Position :

The purist will tell you that a single position line cannot locate your position, and is, therefore, of little value. The practical, experienced navigator, on the other hand, learns to employ any shred of evidence that will assist him in maintaining a secure position and avoiding disaster.

Occasionally you will be thankful for a single position line which will afford you an opportunity to improve your position. The procedure is quite simple. From your DR position drop a perpendicular onto the position line. The point at which this perpendicular line crosses the position line is your improved position.

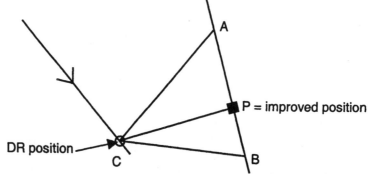

Assuming that the position line is reasonably accurate, the improved position may be shown to be closer to your actual position than the DR position.

Example:

Let A be your actual position. Then P is clearly closer to A than the DR position, since the hypotenuse CA is always longer than either of the other two sides of a right-angled triangle.

Similarly, if B is your actual position, P is closer to B, than C.

The Interception:

The procedure of intercepting another vessel on a different course and at a different speed.

1.   Determine your position and that of your target at a given time, at which you will begin the approach.
2.   Calculate and/or plot the bearing and distance between these two positions.
3.   Set off from your position a line on the reciprocal of the target's course, and measure its length as a proportion of its speed.
4.   With your pair of compasses place the point at the end of the reciprocal, and with a radius proportional to your speed, cut the bearing line.   Join these two points.
5.   This line will give you the course to intercept and the distance along the bearing line will be proportional to your rate of closure.
6.   Divide the original distance by the rate of closure to determine the time of interception.

Note:

If there is a known set and drift, and it is assumed that both vessels are influenced equally by it, then it may be ignored.   However, to determine the coordinates at the time of interception the set and drift must then be applied for the time of the run.

Example:

Let us assume that the target is on a bearing of 035° T. at 60 n.m.
The target's course = 300° T., and its speed = 6 knots.
Your speed = 12 knots.

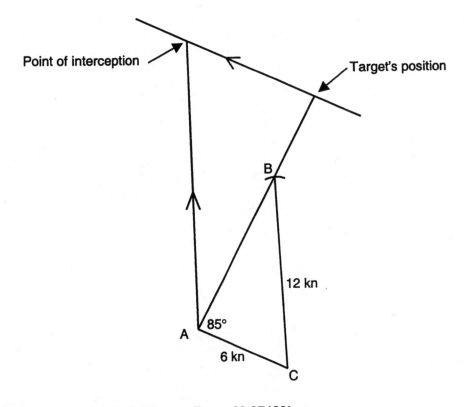

Sin B / 6 = Sin 85° / 12   B = 29.87420°

Then course to interception = 005.l° T.
Angle C = 65.1°
      AB / Sin 65.l° = 12 / Sin 85°
      AB = 10.926

Therefore rate of approach = 10.9 knots.

Time to interception = 5 hours - 29 mins. - 29 sec. (60 ÷ 10.926)

Sea Rescue:

In this day of helicopters and fast Coast Guard cutters it is unlikely that we might be called upon to carry out a search pattern for a missing vessel or life-raft. However it is always wise to err on the side of safety and to be prepared for such an emergency. Furthermore, it is always gratifying to enjoy this sense of competence,

There are many types of search pattern available to us, but one of the easiest to perform is the time-honored "square search". In this procedure we sail to the probable site of the missing craft, allowing for set and drift effect, and remembering that a life raft without a keel is disproportionately affected by the wind. On the way there determine a reasonable visibility distance to avoid over-looking small objects, especially a swimmer. This distance will be improved with binoculars and a flying bridge. If possible determine the set and drift in the area.

On reaching the search commencement point it facilitates your computations to begin the first leg directly into or with the current. The first 2 legs will be equal in length to twice the visibility distance you selected. The next pair will be 4 x , and the 5th. and 6th. legs 6 x, etc.

Knowing the set and drift you can quickly determine the courses and times on each leg, thus, in effect, pre-planning your search. For example let us assume that your speed is 8 knots and the set and drift is 270° / 1.5 knots. The course steered into the current is clearly 090°, and down current 270°, and the ground speeds will be 6.5 knots and 9.5 knots. respectively. On those courses at right angles to the current, by simple triangulation, the drift and the ground speed may be quickly determined. In this example they are 10.8° and 7.9 knots.

From this data set up a table showing the successive courses and the times on each leg, assuming a visibility distance of 0.5 n.m.

Leg	Co. to steer.	Speed	Distance	Time
1.	090.0°	6.5 kn	1 n.m	9 m. 14 s
2.	010.8°	7.9	1	7 m. 36 s.
3.	270.0°	9.5	2	12 m. 38 s
4.	169.2°	7.9	2	15 m. 11 s
			etc.	

The same courses and speeds will be repeated for further legs. Only the distances and times will increase.

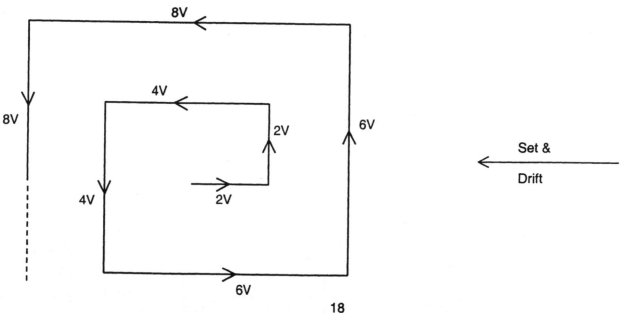

18

The advantage of this method is that the search is a true square plot, and represents your track or ground plot, i.e. your actual position.

Also, since it is a square plot, all the turning points on the first 4 legs are multiples of the relative bearings of 45° from the commencement point. That is the triangles formed are all right-angled isosceles triangles, which means that the successive distances of these turning points from the commencement point are 1.4 x the length of either of the other two sides. Hence, when you finally spot the craft or swimmer, you can quickly identify the true position of the rescue, or the coordinates if further aid is needed.

TIME

Time.

In these hi-tech days, when time is determined by the highly sophisticated measurement of atomic decay, time at sea is no longer measured by burning candles or the sand hour-glass. Fortunately today we have reliable time-pieces employing the quartz crystal, which together with the radio time signals from the National Bureau of Standards assure the modern navigator of a reliable measurement of time. Nevertheless, we are still dependent upon time measurement derived from the daily transit of the Sun about the Earth. Such time is termed "apparent" or true time, which , however, is not a constant rate due, primarily, to the ellipticity of the Earth's orbit about the Sun. The only practical application of this form of time is at local apparent noon (LAT), when the Sun is on our meridian and permits us to determine latitude.

Due to the inconstancy of solar time, an averaged rate has been computed which is constant and lends itself to measurement by our mechanical timepieces. This is termed "Mean Time", which measures a constant rate about a circular orbit, coinciding with the elliptical orbit of solar time at the Vernal and Autumnal equinoxes. When the Earth is passing through apogee and perigee in its orbit about the Sun, i.e. at the Solstices (approx. June and Dec.. 22nd.) there is a significant discrepancy between the Mean and Solar times termed the "Equation of Time" which is recorded in the Nautical Almanac.

Mean time is usually referred to a given meridian. Greenwich Mean time (GMT) is the one preferred by long-range navigators, whereas most of us navigating in local waters prefer local mean time (LMT).

Since LMT is limited to a specific meridian, meridians only a short distance away will have a different time. To avoid this inconvenience the Earth is divided into 15 degree zones within which the time is standardized to the central meridian, hence the name local standard time. Each zone is given a zone description (ZD) which is plus (+) for those West of Greenwich, and minus (-) for those to the East.

E.g. Washington D.C. is in zone +5, therefore at local noon in D.C. the time at Greenwich = 1200 + 5 = 1700 GMT.

Several amendments are made to these zones through-out the world to accommodate certain jurisdictions. These changes are listed in the Nautical almanac.

The Equation of Time, which never exceeds 16 minutes and 25 seconds is listed in the Nautical Almanac for 00 hours (i.e. 180 degrees, the lower branch or anti-meridian of Greenwich) and 12 hours (the Greenwich meridian, 0 degrees). Intermediate points may be interpolated. The time of Meridional passage is given to the nearest minute.

Chapter 2.

This chapter deals with spherical trigonometry and its applications to long-range navigation with great circle sailings, the vertex and its uses, Napier's rules, composite sailing and the conversion angle.

Spherical Trigonometry.

The spherical triangle consists of 3 arcs of great circles on the Earth's surface.

A great circle is a circle on the surface of the Earth the plane of which passes through the center of the Earth.

Thus the Equator and all the meridians are great circles.   Other great circles which are oblique to the Equator and the meridians will lie half in the Northern hemisphere and half in the Southern.
These are the greatest circles that can be drawn on the Earth, and each great circle bisects any other great circle.

Every oblique great circle reaches a maximum latitude in each hemisphere.   This point is called the vertex.   These vertices are numerically equal in latitude but of different sign, i.e. one North and the other South.

Each of these great circles is bisected by the Equator, and the points of intersection are 180° apart, and 90° removed from the longitude of the vertices.   The angle subtended between the great circle and the Equator is numerically equal to the latitude of the vertex.

An arc of a great circle is measured in degrees and minutes, which represents the angle subtended at the center of the Earth by that arc.   The shortest distance between 2 points on the Earth's surface is the great circle arc between them.   The rhumb line distance between them is somewhat longer.

If a great circle is plotted on a Mercator chart it forms a sine curve with a maximum convex to the nearer or elevated pole.

On a gnomonic chart a great circle is a straight line.

A great circle plotted on any chart crosses each successive meridian at a different angle, unlike the rhumb line.

The navigational triangle is composed of 3 arcs of great circles.   The arc between the origin and the elevated pole is termed Co-lat$_1$, i.e., the complement of the latitude.   A similar Co-lat$_2$ for the arc from the destination to the elevated pole.   And the arc representing the great circle track between the origin and the destination.   The apical angle at the pole between the 2 meridians is the DLo.

Solution of the Navigational Triangle:

To determine the distance from the origin to the destination:

$$\text{Cos D} = \text{Cos Co-lat}_1 \times \text{Cos Co-lat}_2 + \text{Sin Co-lat}_1 \times \text{Sin Co-lat}_2 \times \text{Cos DLo.}$$

To determine the initial course angle:

$$\text{Sin C} = \frac{\text{Sin Co-lat}_2 \times \text{Sin DLo}}{\text{Sin D}}$$

Similarly, the course angle into the destination is obtained by:

$$\text{Sin B} = \frac{\text{Sin Co-lat}_1 \times \text{Sin DLo}}{\text{Sin D}}$$

Note:

The course angle is the angle within the triangle, and not necessarily the true course, which may be the supplement of the course angle.

Example:

Determine the initial Course and distance between C (40° N., 75° W.) and B (50° N., 20° W.)

$\text{Co-lat}_1 = 50°$ $\text{Co-lat}_2 = 40°$ $\text{DLo.} = 55°$

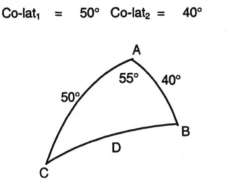

$\text{Cos D} = \text{Cos } 50° \times \text{Cos } 40° + \text{Sin } 50° \times \text{Sin } 40° \times \text{Cos } 55°$
$\quad\quad\text{D} = 39.20991° = 2352.6 \text{ n.m.}$

$\text{Sin C} = \dfrac{\text{Sin } 40° \times \text{Sin } 55°}{\text{Sin } 39.20991°} = 56.4° \quad \text{i.e } 056.4° \text{ T.}$

$\text{Sin B} = \dfrac{\text{Sin } 50° \times \text{Sin } 55°}{\text{Sin } 39.20991°} = 83.04147° \quad \text{i.e. } 096.9° \text{ T.}$

Not infrequently the solution of the navigational triangle involves one or both latitudes in the Southern Hemisphere.

Most navigators prefer not to use the South pole as the elevated pole, but work with the North pole and sides which are greater than 90°. The mathematics are the same but remember that you will often encounter negative numbers which merely represent the second quadrant ratios.

Example:

Determine the initial course and distance from C (20° N., 10° W,) to B (20° S., 70° W.)

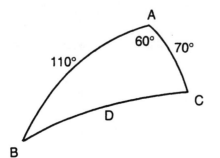

$$\text{Cos D} = \text{Cos } 70° \times \text{Cos } 110° + \text{Sin } 70° \times \text{Sin } 110° \times \text{Cos } 60°$$

$$D = 71.06269° = 4263.8 \text{ n.m.}$$

$$\text{Sin C} = \frac{\text{Sin } 110° \times \text{Sin } 60°}{\text{Sin } 71.06269°}$$

$$C = 59.35766°$$

But looking at the diagram this course is not possible.
Therefore, find its supplement: $180° - 59.35766° = 120.6°$ course angle.
Therefore, the true course to steer $= 239.4°$ T $(360° - 120.6°)$.

It is evident that when solving for the Sine there can be 2 possible answers, which are the supplements of one another.

Try this on your calculator:

Enter 30°. Press sine and record the answer. Clear. Now enter 150°. Press sine. The answer is the same. Now press 2nd. and then arc-sine $(\sin^{-1})$. You should recover 30° instead of the original 150°   This is because these two supplements, in different quadrants, are positive ratios (see the quadrant diagram) which are equal and the calculator cannot discriminate between them.   In the example given it is quite clear which is the correct answer, however, when the initial course angle is close to 90° it is often quite difficult to determine the correct answer.   Fortunately, we have an alternative solution which is not ambiguous - the Haversine.

Haversines:

The versine of an angle is equal to $1 - \text{Cos A}$.

Half a versine or a haversine $= \dfrac{1 - \text{Cos A}}{2}$

There are tables in Bowditch which give these ratios, however, it is quicker and simpler to use your calculator.

Example:

What is the haversine of 30°?

Enter 1, then the minus sign, followed by Cos 30°, strike = ,
then divide by 2 $=$ 0.066987

What is the angle whose haversine is 0.066987 ?

Enter 0.066987 and multiply by 2, $=$ subtract 1 $=$ then change the sign (+ / -). Press 2nd., then arc cos ($\cos^{-1}$) , $=$; answer $= 30°$

The navigational triangle may be solved with the following formulae:

$$\text{Hav D} = \text{Hav DLo} \times \text{Sin Co-lat}_1 \times \text{Sin Co-lat}_2 + \text{Hav (Co-lat}_1 \sim \text{Co-lat}_2)$$

The symbol $\sim$ means "difference between" - simply subtract the smaller from the larger --- i.e. no negative numbers.

$$\text{Hav C} = \text{csc Co-lat}_1 \times \text{csc D} \times [\text{ Hav Co-lat}_2 - \text{Hav (D} \sim \text{Co-lat}_1)]$$

Note:

To obtain the cosecant of a number enter the number, press Sine, then invert (press "1 / x") = cosecant.

Similarly for secant, press cosine rather than sine before obtaining its reciprocal.

Example:

Determine the initial course and distance from C (20° N., 10° W.) to B (20° S., 70° W.)

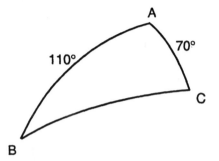

Hav D = Hav 60° x Sin 70° x Sin 110° + Hav 40° =   0.337733
    D = 71.062722°

Hav C = csc 70° x csc 71.062722° x (Hav 110° - Hav 1.062722)
       =   0.754839
    C = 120.64235° = the initial course angle.
    Therefore the course to steer = 239.4° T.

Frequently Used Terms:

1. Two angles whose sum = 90° are complements.
2. Two angles whose sum = 180° are supplements.
3. Two angles whose sum = 360° are explements.

When solving questions involving the spherical triangle we determine one or more of the angles within the triangle. The polar angle is the DLo. or "t" angle, whereas the other two angles are referred to as the "course angles". These angles are not necessarily the navigational courses to be steered (Cn).

If the initial course angle (C) is in the first quadrant, then the course angle = the true course to be steered. This is also true if the course angle is in the second quadrant as derived by the haversine formula.   However, if the Sine law is used the derived angle is always in the first quadrant, and it is left to the navigator to derive the second quadrant value, which is its supplement, i.e. 180° - C.

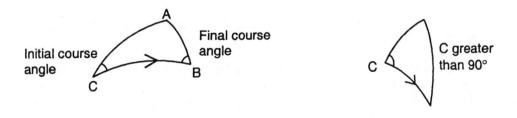

Use of the Sine law always has 2 possible solutions.   If in doubt check with the haversine formula.

Determination of the final course to steer is always the supplement of the derived value in the first and second quadrants.

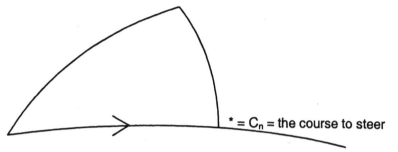

* = C_n = the course to steer

In the 3rd. and 4th. quadrants the initial course angle will be less than 180°, while the navigational course to steer will be its explement, (360° – C)

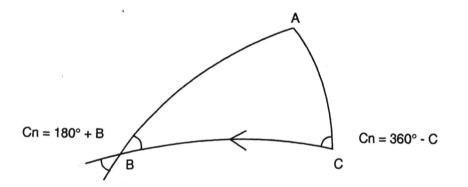

$C_n = 180° + B$ $C_n = 360° - C$

The final course to steer in the 3rd and 4th quadrants = 180° + B.

The Vertex:

This is the highest point of latitude to which the great circle rises, both North and South. Your great circle track is a small arc of this great circle.

The latitude of this vertex (L_v) is always numerically equal to or greater than the latitude of your origin (L_1) and lies towards your destination (L_2) if the initial course is less than 90°.   However, if the course angle is greater than 90° then the nearer vertex lies in the opposite direction. In each case the nearer vertex has the same latitude name (N. or S.) as L_1.

To determine the latitude of the vertex:

Cos L_v = Cos L_1 x Sin C.

where C = the initial course angle, not necessarily the course steered.

To determine the longitude of the vertex:

Sin DLov = Cos C x Csc L_v

To determine the distance from your origin to the vertex:

Sin D_v = Cos L_1 x Sin DLo_v

Example:

You leave 35° N., 70° W. for 50° N., 20° W.   Determine the coordinates of the vertex.

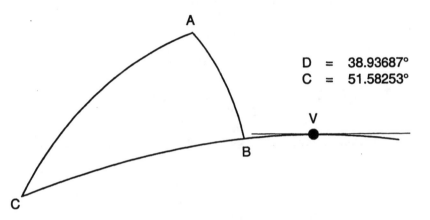

D = 38.93687°
C = 51.58253°

Vertex:

$$Cos\ L_v\ =\ Cos\ 35°\ x\ Sin\ 51.58253°$$
$$L_v\ =\ 50.07316°\ =\ 50°\ 04.4'\ N$$

$$Sin\ DLo_v\ =\ Cos\ 51.58253°\ x\ Csc\ 50.07316°$$
$$DLo_v\ =\ 54.12479°\ =\ 54°\ 07.5'$$

Since C is less than 90° the vertex lies towards L_2, i.e. to the East.   In fact, beyond L_2

$$70°\ W\ -\ 54.12479°\ =\ 15.87521°\ =\ 15°\ 52,5'\ W$$

Therefore the coordinates of the vertex are:

54°04.4' N., 15°52.5' W.

The distance from the origin to the vertex:

$$Sin\ D_v\ =\ Cos\ 35°\ x\ Sin\ 54.12479°$$
$$D_v\ =\ 41.58689°\ =\ 2495.2\ n.m.$$

The most useful application of the vertex is to determine the turning points (T.P's) on the great circle track, so that you can sail a series of rhumb line courses to approximate the great circle.   You may select these T.P's by longitude or distance.

(a) T.P by longitude:

At what Lat$_x$ is your T.P. at longitude 60° W in the previous example ?

$$\text{Tan } L_x = \text{Cos DLo}_{vx} \times \text{Tan } L_v$$
$$= \text{Cos } (60° - 15.87521°) \times \text{Tan } 50.07316°$$
$$L_x = 40.61944° = 40°37.2' \text{ N.}$$

(b) T.P. by distance, e.g. every 300 n.m's (5° of arc):

$$\text{Csc } L_x = \text{Csc } L_v \times \text{Sec } D_{vx}$$
$$= \text{Csc } 50.07316° \times \text{Sec } (41.58689° - 5°)$$
$$L_x = 38.00694° = 38° 00.4'\text{N.}$$

$$\text{Csc DLo}_{vx} = \text{Csc } D_{vx} / \text{Sec } L_x$$
$$= \text{Csc } 36.58689° / \text{Sec } 38.00694°$$
$$\text{DLo}_{vx} = 49.15299°$$

$$\text{Then T.P300} = 15.87521° + 49.15299°$$
$$= 65.02820° = 65° 01.7' \text{ W}$$

$$\text{Therefore TP300} = 38° 00.4' \text{ N., } 65° 01.7' \text{ W}$$

Example:

You depart Taipei (25° N., 122° E.) for Lima (12° S., 77° W)   Determine the coordinates of the vertex and the longitude at which you cross the Equator.

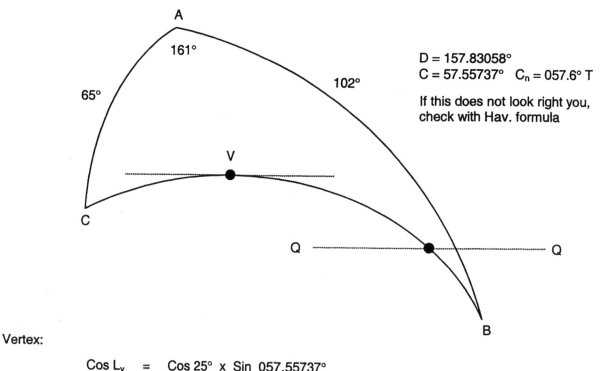

D = 157.83058°
C = 57.55737°   C$_n$ = 057.6° T

If this does not look right you, check with Hav. formula

Vertex:

$$\text{Cos } L_v = \text{Cos } 25° \times \text{Sin } 057.55737°$$
$$L_v = 40.10551° = 40° 06.3' \text{ N}$$

$$\text{Sin DLo}_v = \text{Cos } 057.55737° \times \text{Csc } 40.10551°$$
$$\text{DLo}_v = 56.38222°$$
Therefore $\text{Lo}_v = 178.38222° \text{ E} = 178° 22.9' \text{ E}$

Then the longitude at which you cross the Equator:

$$178° 22.9' \text{ E} + 90° = 91° 37.1' \text{ W}$$

Example:

You depart 30° N., 70° W for 30° S., 20° W.   Determine the coordinates of the vertex, the longitude at which you cross the Equator and your course at that point.

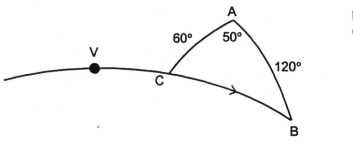

D = 76.57981°
C = 136.99693°

$$\text{Cos L}_v = \text{Cos } 30° \times \text{Sin } 136.99693°$$
$$\text{L}_v = 53.79601° = 53° 47.8'\text{N.}$$

$$\text{Sin DLo}_v = \text{Cos } 136.99693° \times \text{Csc } 53.79601°$$
$$\text{DLo}_v = -65.000003°$$
$$\text{Lo}_v = 135.00000° \text{ W} = 135° 00.0'\text{W.}$$

Since the great circle crosses the Equator 90° from the vertex, the crossing point will be:

$$135.00000° - 90° = 45° 00.0' \text{ W}$$

Since the angle between the great circle track and the Equator equals the latitude of the vertex:

Then the course angle = 90° – 53.8° = 36.2°
Therefore, the course steered at crossing = 143.8° T

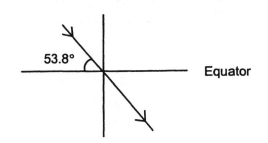

Napier's Rules:

In the solution of spherical triangles special cases exist where one angle or side is equal to 90°.   These cases may be solved by Napier's rules.

Here we begin with a circle and a vertical dropped from the center to a point on the circumference at its lowest point. (6 o'clock).   This point represents the right angle or the right side.   The circle is then divided into 5 equal segments.   Into these segments, beginning at the vertical line and moving clockwise, the parts of the triangle are placed, the sides and angles alternating.   The segments adjacent to the vertical line take the values of the sides or angles unchanged, but the others are labelled as the complements of their respective angles or sides.

Note that the sides or angles added to its complement = 90°,   e.g. 60° + 30° = 90° and 120° + (−30°) = 90°

The Right Spherical triangle:

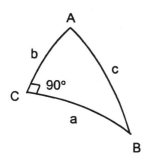

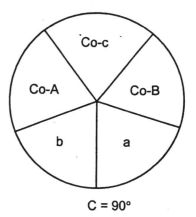

Here the rules are:

The Sine of a middle part equals the product of:

       1.   the tangents of the adjacent parts.

       2.   the cosines of the opposite parts.

Under these circumstances the following rules apply:

    a.   an oblique angle (not a right angle), and its opposite side lie in the same quadrant.

    b.   the hypotenuse, c, (opposite the 90° angle C) is less than 90° when the other two sides, a and b, are in the same quadrant; and more than 90° when they are in different quadrants.

Example:

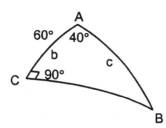

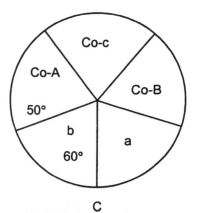

28

$$\text{Sin Co-B} \quad = \quad \text{Cos } 50° \text{ x Cos } 60°$$
$$\text{Co-B} \quad = \quad 18.74724° \qquad \text{Then B} \quad = \quad 71.25276°$$

$$\text{Sin Co-c} \quad = \quad \text{Tan } 50° \text{ x Tan } 18.74724°$$
$$\text{Co-c} \quad = \quad 23.85866° \qquad \text{Then c} = 66.14134°$$

$$\text{Sin a} \quad = \quad \text{Tan } 60° \text{ x Tan } 18.74724°$$
$$\text{a} \quad = \quad 36.00522°$$

or

$$\text{Sin a} \quad = \quad \text{Cos } 50° \text{ x Cos } 23.85866°$$
$$\text{a} \quad = \quad 36.00521°$$

Example:

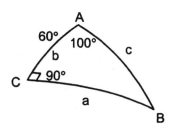
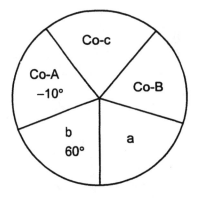

$$\text{Sin Co-B} \quad = \quad \text{Cos -10° x Cos } 60°$$
$$\text{Co-B} \quad = \quad 29.49870°$$
$$\text{B} \quad = \quad 60.501295°$$

$$\text{Sin Co-c} \quad = \quad \text{Tan -10° x Tan } 29.49870°$$
$$\text{Co-c} \quad = \quad -5.72510°$$
$$\text{c} \quad = \quad 95.72510°$$

$$\text{Sin a} \quad = \quad \text{Tan } 60° \text{ x Tan } 29.49870°$$
$$\text{a} \quad = \quad 78.491558° \text{ or } 101.50844°$$

Note: A + Co-A = 90°
100° + (−10°) = 90°

See rule a: since A = 100°, therefore a = 101.50844°

Rule b confirms the value of c in the 2nd quadrant.

The Quadrantal Spherical Triangle:

That is one side = 90°

A similar circle is constructed, but the vertical starting line represents the 90° side. The other segments are again filled by moving clockwise from the 90° line. The two adjacent angles remain unchanged, but all the others are the complements of their values in the triangle.

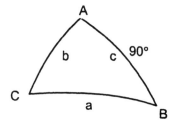

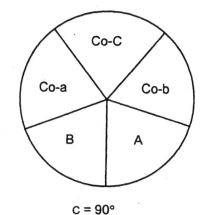

c = 90°

The primary rules for solution are as before:

The Sine of the middle part equal the product of:

    a.   the tangents of the adjacent parts.

    b.   the cosines of the opposite parts.

The interpretive rules are somewhat different from those for the right spherical triangle.

    1.   An oblique angle and the side opposite are in the same quadrant.

    2.   Angle C, the angle opposite the 90° side, is greater than 90° when A and B are in the same quadrant, and less than 90° when A and B are in different quadrants.

Note: If the above rule 2 requires an answer greater than 90°, but its solution is less than 90°, determine the supplement by subtracting it from 180°

Example:

Let c = 90°, A = 50°, and b = 60°   Solve for a, B and C.

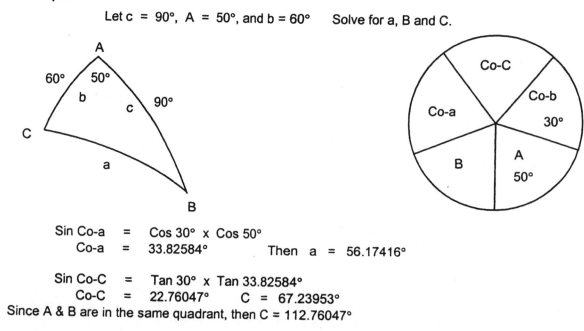

$$\text{Sin Co-a} = \text{Cos } 30° \times \text{Cos } 50°$$
$$\text{Co-a} = 33.82584°\qquad \text{Then}\quad a = 56.17416°$$

$$\text{Sin Co-C} = \text{Tan } 30° \times \text{Tan } 33.82584°$$
$$\text{Co-C} = 22.76047°\qquad C = 67.23953°$$

Since A & B are in the same quadrant, then C = 112.76047°

$$\text{Sin B} = \text{Tan } 33.82584° \times \text{Tan } 50°$$
$$\text{B} = 52.99548°$$

or
$$\text{Sin B} = \text{Cos } 30° \times \text{Cos } 22.76047°$$
$$\text{B} = 52.99550°$$

Note: the difference between the two determinations of B is due to rounding off.

The quadrantal spherical triangle is particularly useful for determining the point at which a great circle track crosses the Equator.

Composite Sailing.

This sailing is a modification of the great circle track in order to avoid a dangerous or obstructive element, e.g. an island, an iceberg, territorial waters etc. This results in a track which begins with a great circle arc, an intervening rhumb line which runs along the selected limiting parallel, and ends with another great circle arc.

This technique can only be used if the vertex lies between the origin and the destination., and assumes that each great circle arc is tangential to the limiting parallel. That is, each great circle arc is at its vertex, and the course angle at that point is 90° (or 270°).

The determination of the longitudes at which you reach and leave the limiting parallel may be obtained by:

$$\text{Cos DLo}_{vx} = \frac{\text{Tan L}_x}{\text{Tan L}_v}$$

where x is the coordinate of the origin and the destination in the successive determinations.

Since the course angle at the limiting parallel = 90° Napier's rules are most easily applied to determine the longitudes of the turning points, the initial and final courses, and the distances run along the arcs. The distance along the limiting parallel is determined by parallel sailing.

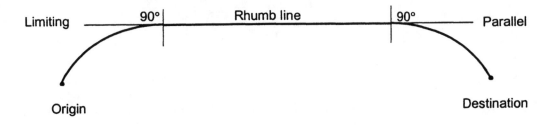

Mercator Sailing:

For those of us who wish to sail a distance of 400 to 600 miles great circle sailing is a little too cumbersome. There is, however, an alternative to mid-latitude sailing which is a little more precise, i.e. Mercator sailing, where the latitude scale is adjusted to compensate for the changing longitudinal scale, termed meridional parts in place of minutes of arc of latitude. These Mer. Parts are tabulated in Bowditch, table 5. These tables are derived from a very intimidating formula, which can be reduced to a useable form without sacrificing a great deal of accuracy.

$$\text{M} = 7915.7 \log \tan (45° + 0.5 \text{ L}) - 23 \text{ Sin L}$$

where M is the number of mer. parts between the Equator and the given latitude L.

Example:

What are the mer. parts to 36° N ?

$$M = 7915.7 \log \tan (45° + 36 / 2) - 23. \sin 36°$$
$$= 2317.98 \ - \ 13.52$$
$$= 2304.5 \quad \text{(Table 5 shows 2304.3)}$$

Example:

Determine the distance between New York (41° N., 74° W) and London (52° N, 0°) by great circle, mid-lat. and Merc. sailings.

a. Great circle sailing

$$D = 49.83018° = 2989.81 \text{ n.m.}$$

b. Mid-lat . sailing:

Diff of L = 11°   DLo = 74° L_m = 46.5°

p = 50.93824°   D = 3126.75 n.m.

c. Merc. sailing:

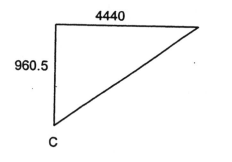

M_{52} = 3646.8 from modified formula.
M_{41} = 2686.3
m   =   960.5   DLo = 74° x 60 = 4440'
Tan C = DLo / m   C = 77.8°
Dist. = d.L x Sec C = 660' x Sec 77.8°
        =   3123.2 n.m.

Chapter 3

This chapter is devoted to the special application of spherical trigonometry to astronomical navigation, with emphasis on sight reductions, star identification, celestial horizon amplitudes, prime vertical sights and meridonial passage of the Sun and the stars. The more practical "moving mer. pass." is also demonstrated.

It will be evident that no tables are employed, and that all calculations are made with basic mathematics. The only exception to this is the inclusion of the long-term almanac, derived from the excellent section in Bowditch, which replaces the Nautical Almanac, until well into the next century.

The Long-Term Almanac (see Appendix C)

The long-term almanac, found in Bowditch, Volume 1, page 1164 in the 1977 edition, is a very useful adjunct to astro-navigation. Although there are small discrepancies, the values obtained rarely exceed 2' of arc, well within the average navigator's accuracy.

Developed from the 1972 data, and the fact that this data repeats itself every 4 years, the tables presented permit the interpolation for any time until well into the next century.

Two sets of time to arc conversion are given. One for the Sun, which changes longitude at the rate of 15° per hour, and the other for Aries and the stars, which change longitude at the rate of 15° 02.46' per hour, or 15.041° per hour.

To determine GHA Aries:
Subtract 1972 from the current year, divide by 4, which gives a whole number and a remainder of 0, 1, 2, or 3. Enter the Aries tables with the month and remainder and obtain the initial figure in degrees and minutes. Then add a correction for the whole number from table D; the day of the month from table E, and the hours, minutes, and seconds in tables F, G, and C respectively. The sum of these 6 figures equals the GHA of Aries.

To determine the SHA and declination of a star:
From the list of stars extract the SHAs and declination for 1972 together with their respective annual corrections, noting their signs. Subtract 1972 from the current year and the decimal representing the month and days. A convenient table is given at the end of the long-term almanac for determining the decimal quickly. Multiply corrections by the years and tenths since 1972, and apply to the respective SHA and declination values, observing their signs.

To determine the GHA and Dec of the Sun:
Subtract 1972 from the current year and divide by 4 to obtain the whole number and remainder. Enter the Sun tables in the segment for the month, and the dates which bracket the time of observation. Extract the "GHA - 175°" before and after the required day, together with the quadrennial corrections.

Extract the declination and its quadrennial corrections by the whole number, and with the results correct the two values for "GHA - 175°" and those for declination.

Obtain the differences between the "GHA – 175 °" values and the declination values. Divide each difference by the number of days separating the two extractions. This is usually 3 days, but may be 4 if the extractions overlap into the next month, or only 2 days if at the leap year. The result equals the daily change.. If the value is falling with time, the correction is minus; and if rising, it is plus. Now multiply this daily rate by the days and tenths since the date of the first extraction, and add or subtract to the earlier values of GHA and dec., to obtain the values for the desired day.

Enter the multiplication tables with the hours, minutes and seconds of GMT (A,B and C respectively) and extract the values to be added to the GHA Sun. Now add 175° to the resultant addition. To this result add or subtract your longitude (E +, W -) to obtain the LHA of the Sun. If this value is greater than 180° subtract from 360° to obtain the "t" angle, which is East.

To determine the altitude and azimuth of the body:

With the GHA of the body, its declination and your DR position obtain your Co-lat1, the bodies Co-dec. and the "t" angle (E or W), and solve the spherical triangle using the same Cosine D and Sine C formulae, as in the navigational triangle, where D = the co-altitude, and C the Azimuth angle. From these values determine the altitude and the true azimuth. The altitude = 90° – co-alt.

If the "t" angle is East, then Z = the azimuth True. If "t" is West, then the True azimuth = 360° - Z.

Example:

Observer at 35° N., 75° W. Dec. of star = 28° 30.0'N. The calculated "t" angle = 44° 40.0' E.

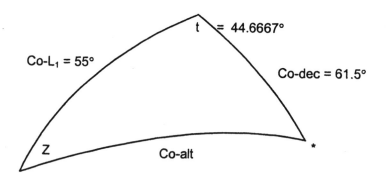

$$\text{Cos co-alt} = \text{Cos } 55° \times \text{Cos } 61.5° + \text{Sin } 55° \times \text{Sin } 61.5° \times \text{Cos } 44.66667°$$

$$\text{co-alt} = 38.21683° \quad \text{Alt.} = 51.78317° = 51° 47.'$$

Using the Sine Law: Z = 87.00077° or 92.9923°
    Hav Z = Csc 55° × Csc 38.21683° × (Hav 61.5° – Hav 16.78317°)
       Z = 87.00004°   Therefore, Az. = 087.0° T

Star Identification.

To identify a star determine your DR. position, the time (GMT) and the altitude and bearing of the body..

To determine the declination:

Cos co-dec = Cos co-L_1 x Cos co-alt. + Sin Co-L_1 x Sin co-alt x Cos Z

Example:

Observer's position = 36° N. 76° W.   Az of * = 056°
Altitude of *   = 31° 30.0'

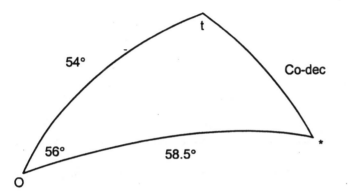

Co-dec = 46.14399°   dec. = 43.85601° =   43° 51.4' N.

Sin "t" = Sin 58.5° x Sin 56°
            Sin 46.14399°   "t" = 78.60546° E

GHA Aries = calculated for time of observation.

"t" = 78° 36.3 E   Then LHA * = 281° 23.7

Long. = 76° 00.0' W   Therefore GHA * = 357° 23.7'

GHA Aries + SHA * = GHA *

Then SHA * = GHA * - GHA Aries.

Identify star with the calculated dec. and SHA.

Meridional Passage

Most amateur navigators learn the principle of the Mer. Pass. by calculating the time of this event for a given fixed position.   In practice however, this rarely is the case.   Almost invariably the Mer. Pass. is determined for a moving vessel, and the navigator simply waits until the body is at an altitudinal zenith, i.e. it is due North or South on your meridian.
An altitude at this point provides a measure of the ship's latitude.   The time of the event in GMT will also provide a measure of your longitude, provided the equation of time is known.   Alternatively the time and place of Mer. Pass. can be pre-calculated.

I am sure we all recall the problem of a cyclist departing point A, at the same time that a car leaves B at their respective speeds. Given the distance between A and B, we determine the time and place at which they meet, by calculating the rate of approach. In the navigational problem we determine the longitude of the body and that of the vessel at a given time prior to Mer. Pass., and calculate the DLo, the vessel's rate of change of longitude per hour. If the body is the Sun its westerly movement is 15° per hour, and if it is a star its rate is 15° 02.5' per hour. The time of Mer. Pass. is the DLo. divided by the rate of approach, which is the sum of the 2 rates, if the vessel is on an easterly course, and the difference if on a westerly course. The new position can now be calculated.

Example:

Let us assume that you have calculated that the Sun crosses the Greenwich meridian at 1156 GMT, and that your DR position at that time is 40° N., 60° W.

If your course is 090° T., at 12 knots, then your rate of change of longitude = 12' / Cos 40° = 15.6649' / hr. = 0.26108° / hr. Therefore, the rate of approach = 15° + 0.26108° = 15.26108° / hr. The time to Mer. Pass. = 60° (Greenwich to your DR position) divided by 15.26108° = 3.93157 hrs. You can now calculate your new DR position at Mer. Pass. for the time = 1156 + 0356 = 1552 GMT.

This, however, is an extreme case, since you rarely have the good fortune to be travelling due East or due West. Far more frequently you are on an oblique course.

Let us assume you are on a course of 050° T at 12 knots. At 40° N latitude your rate of change of longitude is:

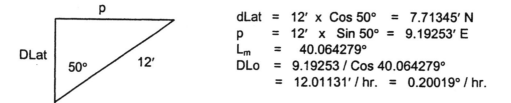

dLat = 12' x Cos 50° = 7.71345' N
p   = 12' x Sin 50° = 9.19253' E
L_m =   40.064279°
DLo = 9.19253 / Cos 40.064279°
     = 12.01131' / hr. = 0.20019° / hr.

Therefore time to Mer. Pass. = 60° / 15.20019° = 3.94732 hrs.

If, however, you had been on a course of 310° T, then the rate of approach would have been 15° − 0.20019° = 14.79981° / hr.

60° DLo / 14.79981° = 4.05411 hrs = 4 hrs. 03 mins. 15 secs

Outline of procedure for determining the time and place of Mer. Pass. for a moving observer:

1. Determine GMT of body on the Greenwich meridian, or any other convenient meridian.

2. Determine your position at that time.

3. Determine your rate of change of longitude per hour.

4. Determine the rate of closure (plus if your course has an Easterly component; minus if it has a Westerly component). Unless your course is due E or W, this will not be equal to your speed in terms of d. long. per hour.

        The Sun's rate   =   900' arc d.long. / hr.
        A star's rate   =   902.46' arc d.long. / hr.

Special Great Circles

I. The Celestial Horizon:

This is the great circle on the celestial sphere, the plane of which passes through the center of the Earth. This horizon lies a smidgen above your visible horizon, and is parallel to it.

When the center of a body lies on the celestial horizon its altitude is exactly 0°, and therefore, the co-altitude is 90°, which allows the use of Napier's rules to calculate the "t" angle and the azimuth angle.

A bearing of the body at this time provides a check on the deviation of your compass, however, since you cannot be sure exactly where the celestial horizon lies, it is customary to take a bearing when the body is on your visible horizon, and apply a small correction derived from table 28 in Bowditch. This correction is rarely as high as 1° for declinations less than 25°, and latitudes between 25° and 50°.

In addition to Napier's rules, the azimuth and "t" angles may also be solved by that formula for spherical triangles in which all 3 sides are known, i.e. co-lat1 , co-dec., and co-alt.

$$\text{Hav "t"} = \frac{\text{Hav co-alt} - \text{Hav (co-lat}_1 \sim \text{co-dec)}}{\text{Sin co-lat}_1 \times \text{Sin co-dec.}}$$

The azimuth is then calculated with the Sine law.   This azimuth angle may also be derived by Table 27, which provides the amplitude, i.e. the angle North or South of due East or due West.   Or by the following formula:

$$\text{Sin amplitude} = \text{Sec latitude} \times \text{Sin declination.}$$

2.   The Prime Vertical Circle:

This is the great circle which begins at your zenith and cuts the celestial horizon at right angles, due East or West of you. This also affords you an opportunity to check your deviation.

Since the azimuth angle, when the body is on the prime vertical, is exactly 90° East or West, Napier's rules may be used to determine the "t" angle.

3.   Having established a "t" angle for the celestial horizon or the prime vertical observations the time to elapse before Mer. Pass. is easily established by dividing the "t" angle (i.e. the DLo. between you and the body) by 15° if the body is the Sun, or by 15.04167° if a star, to determine the hours before Mer. Pass.

More on the Prime Vertical:

This great circle runs from your zenith, due East or West, cutting your celestial horizon at right angles.

The following rules apply:

1. A body with a declination of opposite name to your latitude crosses your prime vertical below the horizon
2. A body with a declination of the same name as your latitude, but numerically greater, does not cross your prime vertical.
3. A body with a declination of the same name as your latitude, but numerically smaller, crosses the prime vertical at some distance from your meridian, first to the East and later to the West. In each case the meridian angle ("t") is equal in each case, and always less than 90°
4. The formulae to be employed are:

$$\text{Cos "t"} = \text{Tan dec.} \times \text{Cot L}$$

$$\text{Sin h (alt)} = \text{Sin dec.} \times \text{Csc L}$$

A sight obtained of a celestial body when it is on your prime vertical can only occur if its declination is numerically less than your latitude, and bears the same name, i.e. its declination lies between your latitude and the Equator.

Calculating the Conversion Angle.

This angle is the difference between a great circle bearing and the rhumb line bearing of a given point.

This calculation is most useful in converting a received radio signal, which follows a great circle from the transmitting station to your position, to the rhumb line which can be plotted on your Mercator chart from the station and gives a position line through your position.

This angle is derived from a simplified formula, which is a practical abbreviation of a more complex, precise mathematical formula:

$$\text{Tan conversion angle} = \text{Sin } L_m \times \text{Tan } 0.5 \text{ DLo}$$

How to apply the Conversion angle to the great circle bearing to obtain the rhumb line bearing:

Since circles plotted on a Mercator chart curve towards the nearer pole, the C.A. will be applied to the great circle bearing with a different sign depending upon which hemisphere you are in, and also according to whether you are East or West of the station transmitting the signal

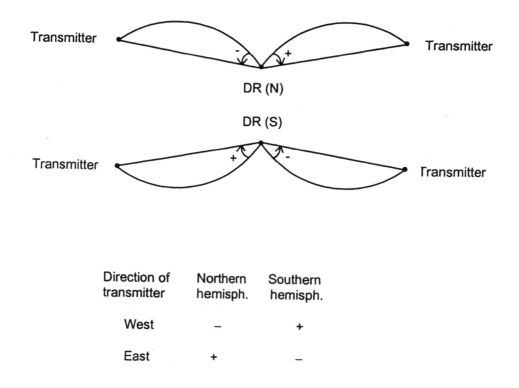

Direction of transmitter	Northern hemisph.	Southern hemisph.
West	–	+
East	+	–

Single Sight Fixes:

It is generally assumed that a fix requires the intersection of two lines of position.   However, there are two exceptions to this assumption, which provide a fix with a single altitude determination at a known time (GMT).

These exceptions are:

    a. Prime vertical sight.
    b. Meridian Passage sight.

Prime Vertical sight:

A sight of a body on your prime vertical provides you with its altitude and the GMT of the observation.

Since a body on the P.V. bears 090° or 270°, the astro-triangle is a right spherical triangle, and lends itself to solution by Napier's rules.   Your latitude and the "t" angle may then be calculated.   When the GHA of the body is determined then your longitude is the GHA of the body plus or minus the "t" angle, depending upon which side of you (E or W) the body lies.

Mer. Pass. Sight:

Since a Mer. Pass. of the Sun or a star has its GHA = to your longitude, then the calculation of the body's GHA from the GMT of the sight will provide your longitude.

A sight taken at Mer. Pass. provides an altitude, which adjusted for the bodies declination will give you your latitude.

Example:

At GMT 0700 on July 4th. the Sun is on the observer's Prime Vertical (East) with an altitude = 42° 04.4' (corrected).

GHA Sun = 283° 56.1'   Dec. = N 22° 54.'

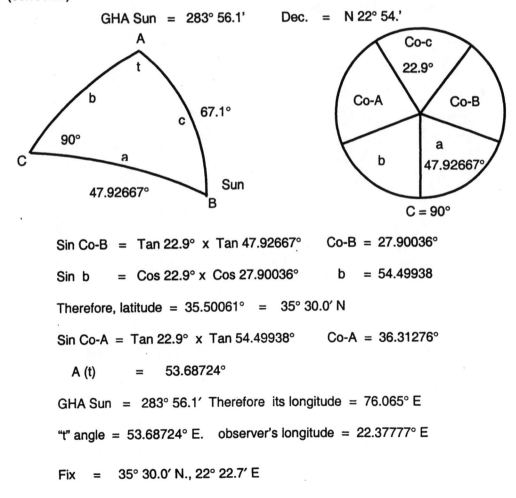

Sin Co-B = Tan 22.9° x Tan 47.92667°   Co-B = 27.90036°

Sin b   = Cos 22.9° x Cos 27.90036°   b = 54.49938

Therefore, latitude = 35.50061° = 35° 30.0' N

Sin Co-A = Tan 22.9° x Tan 54.49938°   Co-A = 36.31276°

  A (t)   =   53.68724°

GHA Sun = 283° 56.1' Therefore its longitude = 76.065° E

"t" angle = 53.68724° E.   observer's longitude = 22.37777° E

Fix =   35° 30.0' N., 22° 22.7' E

39

Example:

At 1700 GMT on July 4th. the Sun is on your meridian to the South, with an altitude = 77° 21.8′ (corrected)

GHA Sun = 73° 55.0′   Dec. = N 22° 51.8′

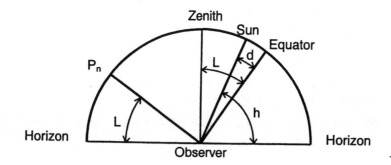

Zenith distance = 90° – h   L = z.d. + dec.

L = (90° – h) + dec. = (90° – 77° 21.8′) + 22° 51.8

Therefore Latitude = 35° 30′N.

Fix: 35° 30.0′ N., 73° 55.0′ W

Practice Questions.

Chapter 1, Section 1:

1. You depart 37° 14.0′ N., 75° 23.0 W, and sail 33 n.m. on a course of 078°T

   What are the coordinates of your destination ?
   (answer: 37° 20.9′ N., 74° 42.4″ W)

2. You are to sail from 24° 17.0′ N., 80° 53.0′ W to 22° 57.0′ N., 86° 33.0′ W.
   Determine the course and distance.
   (answer: 255.6° T., 321.6 n.m.)

3. You leave 01° 15.0′ S., 04° 26.0′ E to sail 100 n.m. on a course of 293° T.
   What are the coordinates of your destination ?
   (answer: 00° 35.9′ S., 02° 53.9′ E.)

4. You are on a course of 123° T.   At 0936 a buoy has a relative bearing of
   333° rel.   At 0956 the buoy bears 267° rel.   Your speed = 7.2 knots.
   How far are you from the buoy at your second bearing ?   What is its true
   bearing ?   (answer: 1.2 n.m., 030° T.)

5. You are on a course of 255° T at 6.6 knots.   At 11-17-30 a lighthouse
   333° T.   At 11-40-15 it bears 014° T.   If the position of the lighthouse is
   13° 23.5′ S., 72° 44.6′ W., what is your fix at 11-40-15?

   (answer: 13° 27.1′ S., 72° 45.5′ W.)

6. You are to sail from the Falkland Islands, 52° 10.0′ S., 58° 43.0′ W, to
   South Georgia, 54° 01.1′ S, 37° 57.4′ W. by mid-lat. sailing, at 9 knots.
   The predicted set and drift = 165° / 2.5 knots.   If the variation is 0°, i.e.
   you are on the Agonic line, and the deviation = 4° W, what is your compass course to
   steer ? If you depart at 0600 LST on Jan. 4th, what is your ETA (GMT) ?

   (answer: track = 098.4° T. , dist. = 756.2 n.m., course = 087.7° C.,

   ETA = Jan. 7th., 15-58-39 GMT.)

Chapter 1, Section 2:

1. You are cruising past a prominent landmark on a course of 045° T, at 8 knots.   At 1001 by
   radar scope the landmark is 1.3 n.m. off your port bow.   At 1016 it is 2.1 n.m. off your port
   quarter.   If the landmark is located at 37° 37.6′ N., 76° 24.1″ W, what is your fix at 1016?
   (answer: 37° 37.9′ N., 76° 21.5″ W.)

2. You are cruising on a course of 270° T. at 8 knots.   At 1114 an object is observed at a relative
   angle of 043° rel.   At 1129 your range-finder notes the object at 1.5 n.m. on your starboard
   quarter.   If the object is at
   37° 36.0′ N., 76° 14.2′ W., what is the fix at 1129 ?

   (answer: 37° 34.6′ N., 76° 14.8′ W.)

3. You are cruising on the Chesapeake Bay.   At 1233 you observe a blip on your radar scope bearing 317° T. at 1.8 n.m.   At 1248 the same blip bears 340° T at 1.3 n.m.   If the blip is Stingray Point Light at 37° 33.6′ N.,76° 16.2′ W., determine your speed, your course and your position at 1248.
(answer: 3.15 knots, 276.9° T., 37° 32.4′ N., 76° 15.6′ W.)

4. You are cruising on a course of 078° T.   You observe a buoy which bears 333° relative. Twenty minutes later the buoy bears 222° rel., and is estimated to be 1.1 n.m. off.   What is your speed?
(answer: 6.786 knots.)

5. You set out to determine your set and drift.   Passing very close to a buoy on a heading of 043° T., you obtain relative bearings of the buoy until the readings are consistent at 175° rel.   Your speed is 6 knots. At exactly 10 minutes after passing the buoy you do a sharp turn onto the course which will place you on a reciprocal track back to the buoy. The time to reach the buoy after the turn is 11 mins.,15 secs.   Determine the set and drift.
        (answer: 341.9° T. / 0.63 knots)

6. You pass close to Buoy A at 8 knots on a course of 123° T.   After 7 mins., the buoy bears 186° rel. At 8 mins. 30 secs. you alter course to 038° T. After 9 minutes on this leg the buoy bears 225° rel.   Determine the set and drift.
(answer: 234.7° T / 0.9 knots.)

7. You are sailing in the Gulf of Alaska on a course of 260° T at 9 knots.   At 1100 LST your GPS reading is 55° 35.1′ N., 143° 17.7′ W. One hour later the GPS reading is 55° 32.4′ N, 143° 31.7′ W.
What is the set and drift ?   (answer: 140° T / 1.5 knots.)

8. You are stationed at Ascension Island, 08° S., 14° W.   You are a ham operator, and at 1035 LST you contact a station at Jakarta, 06° S., 107° E. One hour later you establish contact with a ham at Hilo, 20° N., 155° W. Determine the LST of each contact.
(answer: Jak. = 1835 LST.   Hilo = 0235 LST.)

9. The Sun crossed the Greenwich meridian at 11-53-30 GMT.   What will be the LST of Mer Pass. at Pitcairn Island, 25° S., 127° W., and at Taipei, 25° N., 122° E. ?
(answer: Pit. = 12-21-30 LST.   Tai. = 11-45-30 LST.

Chapter 2:

1. Determine the initial course, distance and final course from Cape Henry, 36° 56.0′ N., 76° 01.0′ W. to Lisbon, 39° 14.2′ N., 09° 23.0′ W.

   (answer: 065.7° T., 3076.3 n.m., 109.8° T.)

2. Determine the same data from Cape Henry to Ascension Island, 08° 11.2′ S., 14° 14.4′ W.
   (answer: 114.4° T., 4393.6 n.m., 132.6° T.)

3. Determine the same data from Cape of Good Hope 34° 15.2′ S., 19° 11.4′ E. to Cape Horn 55° 17.1′ S., 68° 39.4′ W.
        (answer: 220.5° T., 3677.5 n.m., 289.7° T)

4. You sail a great circle track from New York, 41° 04.7′ N., 73° 58.6′ W to Land's End, 50° 05.4′ N, 06° 55.9′ W. Determine the coordinates of the vertex and the distance to this point.
   (answer: 51° 52.0′ N; 27° 09.4′ W; 2000.7 n.m.)

5. You sail from Perth 31° 17.2′ S., 115° 57.9′ E to Durban 29° 49.6′ S., 31° 10.4′ E. Determine the coordinates of the vertex and the distance to this point. Also determine the turning points at 20° and 30° DLo. from the vertex. Determine the turning point at 1200 n.m. from the vertex.
   (answers: Vertex = 38° 39.6′ S., 75° 23.7′ E. Turning points
   L_x20 = 36° 56.1′ S., L_x30 = 34° 42.9′ S, L_x1200 = 35° 56.8′ S., 50° 24.3′ E.)

6 You sail from Cape Henry 36° 56.0′ N., 76° 01.0′ W to Cape Town 34° 16.4′ S., 17° 59.1′ E. At what longitude do you cross the Equator ?
   Compute by Napier's rules and confirm by the vertex method. Does this confirm that the angle between the course and the Equator = the latitude of the vertex ?
   (answer: 26° 00.4′ W.)

7. You leave Cape Henry 36° 56.0′ N., 76° 01.0′ W on a great circle track, the initial course of which is 090° T. At what latitude will you cross the 45° W meridian ? (answer : 32° 47.5″N.)

8. You leave Mogadishu 02° 00.0′ N., 45° 00.0′ E to sail a great circle track. You pass through 35° of longitude and sail 2114 n.m. Where is your landfall ? (answer: Colombo, Sri Lanka 06° 69.7′ N., 80° E.)

9. You leave position "A" for a destination "B". Your initial course = 027.8° T. You pass through 58° of longitude. Your final course = 072.8° T. Identify your destination if "A" is Miami 26° N., 80° W. (answer: Reykjavick 63° 58.5′ N., 22° W.)

Chapter 3:

1. You are on Fiji at 19° 55.5′ S., 178° 03.′ E. You calculate the altitude and azimuth of the Sun at 1015 LST on Dec. 25th. 2000. Determine whether these findings will be exactly the same 1 year later.
   (answers: 2000 = Alt. 63° 37.0′, Az. 102° T
   2001 = Alt. 63° 38.8′, Az. 102.6735° T)

2. You depart Taipei, 25° N., 122° E. for Vancouver, 49° N., 123° W, on June 11th, 2000, at 1000 LST at 12 knots, following an approximate great circle track. On June 14th at 13-20-00 GMT you observe a star bearing 075.2° T, at an altitude = 57° 38.6′, per bubble sextant. Identify the star. (answer: Vega.)

3. While sailing in the Indian Ocean you are uncertain of your position. At 0100 GMT on Feb. 4th. 2001 you note the star Regulus bearing 287.1° T, at an altitude, by bubble sextant, = 30° 41.4′. Determine your position.
   (answer: 05° 01.3′ S, 60° 01.1′ E)

At essentially the same time an unknown star bears 124.7° T., at an altitude = 47° 48.1′.
Determine the unknown star.
(answer: Antares.)

4. Determine the GMT of Mer. Pass. of the Sun at Okinawa, 26° N, 128° E., on May 17th, 2002.
  (answer: 03-24-14 GMT)

5. You depart Ascension Island, 08° S., 14° W., on an initial course of a great circle track of
   145° T., at 8 knots, departing at 0600 GMT on June 7th. 1999. Determine the time and place
   of the Mer. Pass. of the Sun on June 15th.
   (answer: 11-48-30 GMT., at 28° 54.9′ S., 02° 50.5′ E.)

6. You depart Cape Town 34° S., 18° E., at 1200 LST on Oct. 17th. 2001, on an initial course
   of 285° T. at 12 knots.   Determine the Mer. Pass. of Canopus on Oct. 22nd.

   (answer: 04-47-08 GMT)

7. On Dec. 22nd. you are sailing on a southerly course.   At 15-59-00 GMTthe Sun, whose
   declination = S 23° 26.5′, casts a shadow of your 20 ft. mast on the foredeck bearing 180°
   T, and measuring 10.625 ft. long.   The Sun had passed the meridian of Greenwich at
   11-59-00 GMT. What is your fix ?
   (answer: 51° 25.3′ S, 60° W = Falkland Isles.)

   8. You are somewhere in the N. Pacific on June 24th. 1999.   At 02-57-09 GMT Sirius
   bears 180° T.   Determine the coordinates of your position if at the time of Mer Pass. Sirius
   has a corrected altitude = 59° 45.2′.
       (answer: 13° 32.0′ N., 145° 09.8′ E. = Guam)

9. You are at 35° N., 75° W. and observe Arcturus to be due West on June 15, 1997.
   Determine the time and altitude of your observation.

      (answer: 05-42-05 GMT., altitude = 34° 58.5′)

10. You are located at 37° N., 77° W. on Jan. 24th. 1997.   Determine the time and azimuth at
    which Aldebaran will be on your Eastern Celestial Horizon.
       (answer: 120-18-32 GMT. Az. = 069.2° T)

    Also determine the time at which it will cross your meridian.
       (answer: Mer. Pass. at 20-26-08 LST.)

Appendix A

Additional Useful Formulae in the Solution of the Spherical Triangle:

1. Three angles: A, B & C:

    $$S = 0.5(A + B + C)$$

    $$\text{Hav } a = \frac{-\text{Cos } S \times \text{Cos}(S - A)}{\text{Sin } B \times \text{Sin } C}$$

    The remaining two sides may now be calculated with the spherical Sine law.

2. Two angles with one opposite side: A, a, B.

    a. $\text{Cot } N = \text{Tan } B \times \text{Cos } a$

    b. $\text{Sin}(C - N) = \dfrac{\text{Cos } A \times \text{Sin } N}{\text{Cos } B}$

    Since this formula is solved by a Sine formula, 2 possible solutions occur, i.e. supplemental values.

    Once C has been established the other 2 sides are calculated by the Sine law.

3. Two angles and the side opposite the unknown angle: A, B, c:

    $$\text{Cos } C = \text{Sin } A \times \text{Sin } B \times \text{Cos } c - \text{Cos } A \times \text{Cos } B$$

    With an angle and its opposite side known (C, c) you can use the Sine law to find the other 2 sides.

4. A known angle and 2 sides, one of which is the side opposite the known angle: A, a, b:

    a. $\text{Tan } M = \text{Tan } A \times \text{Cos } b.$

    b. $\text{Sin}(C + M) = \dfrac{\text{Sin } M \times \text{Tan } b}{\text{Tan } a}$

    Once C is established, the remainder of the unknowns may be calculated by the Sine law.

5. One known angle and two sides, neither of which is the side opposite the given angle; i.e. 2 sides and the included angle: C, a, b.

    $$\text{Hav } c = \text{Hav}(a \sim b) + \text{Sin } a \times \text{Sin } b \times \text{Hav } C$$

6. Three known sides, but no angles: a, b, c:

    $$\text{Hav } A = \frac{\text{Hav } a - \text{Hav}(b \sim c)}{\text{Sin } b \times \text{Sin } c}$$

Remember:

Always bear in mind that whenever a solution is obtained which is a function of the Sine, you must consider the possibility that the true answer is the supplement of the value obtained.

45

Appendix B

Graphic Determination of Trig Function

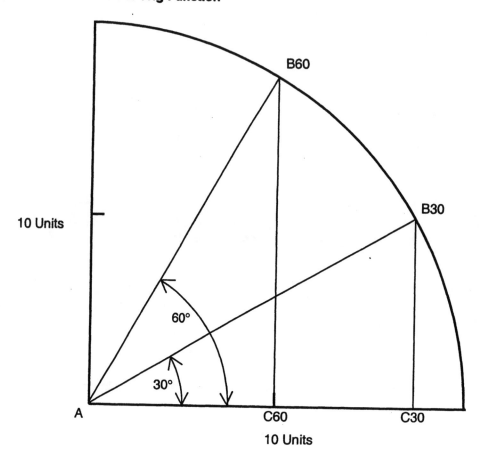

1. Draw quadrant with radius of 10 units (cm or inches).

2. Draw radii for given angles by protractor, or bisect.

3. Measure length of AC and BC:

    Sine = BC/radius
    Cosine = AC/radii
    Tan = Sine/Cosine

Appendix C

The Long-Term Almanac.

Extracted from the American Practical Navigator

by

Nathaniel Bowditch.

1977 Edition

Using the Long Term Almanac

1. Determining the GHA and Declination of the Sun.

    Example:
    Determine the GHA of the Sun, and is declination at 10-30-40GMT on July 5th, 2001.

2001-1972	=	29 ÷ 4 = 7 with a remainder of 1.
GHA – 175, 7/4 (1R)	=	3° 56.6′ Corrn = – 0.07′
GHA – 175, 7/7 (1R)	=	3° 48.8′ Corrn = – 0.05′

3° 56.6′ - (0.07 x 7)	=	3° 56.1′
3° 48.8′ - (0.05 x 7)	=	3° 48.4′
3 day change	=	– 7.7′
1 day change	=	– 2.6′

Change since 0000 July 4th   =   – 2.6′ x 1.4 = – 0.4′

Then GHA – 175	=	3° 56.1′ – 0.4′	=	3° 55.7′
GHA	=	3° 55.7′ + 175°	=	178° 55.7′
A (10 hours)			=	150°
B (30 mins)			=	7° 30.0′
C (40 secs)			=	10.0′
Gha Sun @ 10-30-40 GMT			=	336° 35.7′

Declination:

7/4 (1)	=	22° 54.7′ N	Corrn	= – 0.31 x 7	=	22° 52.5′ N
7/7 (1)	=	22° 37.6′ N	Corrn	= – 0.34 x7	=	22° 35.2′ N
3 day change					=	– 17.3′
1 day change					=	– 5.8′

Change since 7/4 =   22° 52.5′ – (5.8′ x 1.4)   =   22° 44.4′ N

If DR position at 10-30-40 GMT on July 5th 2001 is at 40° 00.0′ N, 31° 00.0′ W, then

LHA Sun   =   336° 35.7′ – 31° 00.0′ W   =   305° 35.7′

"t" angle   =   54° 24.3′ E   54.405° E

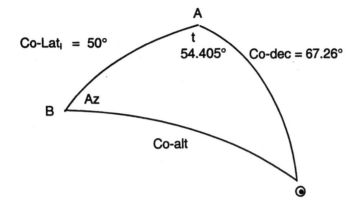

Using the Cosine formula:

Cos Co-alt = Cos 50° x Cos 67.26° + Sin 50° x Sin 67.26° x Cos 54.405°

Co-alt = 48.72388°

Alt = 41.27612° = 41° 16.6′

Using the Sine formula:

$$\frac{\text{Sin Az}}{\text{Sin } 67.26°} = \frac{\text{Sin } 54.405°}{\text{Sin } 48.72388°}$$

Az = 86.26634° or 93.73366°

Az = 93.73366° checked by haversine formula

Therefore AZ of Sun = 93.7° T

2. Determining the GHA of Aries, and the SHA and declination of a given body.

Example:
Determine the GHA and declination of Arcturus at 01-40-30 GMT on Aug. 12th 2000.

GHA Aries:

Aug / 0R	=	308° 42.8′
D (7)	=	12.9′
E (12)	=	11° 49.7′
F (01)	=	15° 02.5′
G (40)	=	10° 01.6′
C (30)	=	7.5′
GHA Aries	=	345° 57.0′

Arcturus:

1972 SHA	=	146° 24.2′	Annual Corrn	=	− 0.68
				x	28.6
					− 19.4
SHA*	=	146° 04.8′			
GHA Aries	=	345° 57.0′			
GHA *	=	492° 01.8′ - 360	=		132° 01.8′

Declination = 19° 19.6′ corrn = 0.31 x 28.6 = -8.9

Then declination in 2000 = 19° 10.7′ N

If DR position @ 01-40-30 GMT on Aug 12th, 2000 is 37° 30.0′ N, 75° 40.0′ W,

Then LHA * = 132° 01.8′ − 75° 40.0′ = 56° 21.8′

"t" = 56° 21.8′ W = 56.36333°

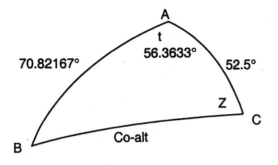

Co-alt	=	52.04429° by Cosine formula
Alt	=	37.95571° = 37° 57.3′
Azimuth angle by Sine rule	=	85.79115° or 94.20885°
Haversine formula reveals	=	94.20885°
Azimuth	=	265.8°

3. Star Identification

Example:
At 02-37-20 GMT on Aug 10th, 2002, in DR position 37° 26.0′ N, 64° 17.6′ W, a bright star bears 281.3° T with an altitude = 78° 45.5′. Identify it.

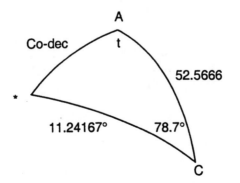

By Cosine formula   Co-dec   =   51.20702°

                          Dec   =   38.79298°   =   38° 47.6′ N

By Sine formula "t" angle   =   14.19791°

                                  =   14° 11.9′ W

             Longitude   =   <u>64° 17.6′ W</u>

      Therefore GHA *   =   78° 29.5′

GHA Aries:

      2002-1972   = 30 ÷ 4 = 7 + remainder of 2

GHA Aries   =   308° 14.2′

     D 7   =   12.9′

     E 10   =   9° 51.4′

     F 02   =   30° 04.9′

     G 37   =   9° 16.5′

     C 20   =   <u>5.0′</u>

                    =   357° 44.9′

GHA Aries + SHA *   =   GHA *

             SHA *   =   GHA * - GHA Aries

                      =   78° 29.5′ – 357° 44.9′

                      =   (78° 29.5′ + 360°) – 357° 44.9′

                      =   438° 29.5′ – 357° 44.9′

                      =   80° 44.6′

Then:

        SHA *   =   80° 44.6′   Dec   =   38° 47.6′ N

        i.e., the star   =   Vega.

APPENDIX H

LONG-TERM ALMANAC

This appendix is intended for use when a more complete almanac is not available It is based principally upon the fact that approximately correct values for the Greenwich hour angle and declination of the sun, and the Greenwich hour angle of Aries, can be obtained from an almanac that is exactly four years out of date. The differences in these values at intervals of exactly four years can be largely removed by applying an average correction to the values obtained from the tables of this appendix. The maximum error in an altitude computed by means of this appendix should not exceed 2ʹ0 for the sun or 1ʹ3 for stars.

This four-year, or quadrennial, correction varies throughout the year for the GHA of the sun (between about plus and minus one-half of a minute) and for the declination of the sun (between about plus and minus three-fourths of a minute). For the GHA of Aries the quadrennial correction is a constant, (+)1ʹ84. The appropriate quadrennial correction is applied once for each full four years which has passed since the base year of the tabulation (1972 in this appendix).

The tabulated values for GHA−175° and declination of the sun and GHA of Aries are given in four columns, labeled 0, 1, 2, and 3. The "0" column contains the data for the leap year in each four-year cycle and the 1, 2, and 3 columns contain data for, respectively, the first, second, and third years following each leap year.

The GHA−175° and declination of the sun are given at intervals of three days throughout the four-year cycle, except for the final days of each month, when the interval varies between one and four days. Linear interpolation is made between entries to obtain data for a given day. Additional corrections to the GHA of the sun of 15° per hour, 15ʹ per minute, and 15ʺ per second are made to obtain the GHA at a given time. Declination of the sun is obtained to sufficient accuracy by linear interpolation alone.

The GHA of Aries is given for each month of the four-year cycle. Additional corrections of 0°59ʹ14 per day, 15°02ʹ5 per hour, 15ʹ per minute, and 15ʺ per second are made to obtain the GHA at a given time.

The SHA and declination of 38 navigational stars are given for the base year, 1972.0. Annual (not quadrennial) corrections are made to these data to obtain the values for a given year and tenth of a year.

A multiplication table is included as an aid in applying corrections to tabulated values.

Sun tables. 1. Subtract 1972 from the year and divide the difference by four, obtaining (a) a whole number, and (b) a remainder. Enter column indicated by remainder (b) and take out values on either side of given time and date.

2. Multiply quadrennial correction for each value by whole number (a) obtained in step 1 and apply to tabulated values plus 175°.

3. Divide difference between corrected values by number of days (usually three) between them to determine daily change.

4. Multiply daily change by number of days and tenths since 0ʰ GMT of earlier tabulated date, and mark correction plus (+) or minus (−) as appropriate.

5. (GHA only.) Enter multiplication table with hours, minutes, and seconds of GMT, and take out corrections A, B, and C, respectively. These are all positive.

6. Apply corrections of steps 4 and 5 to corrected earlier values of step 2.

Example.—Find GHA and declination of sun at GMT 17ʰ13ᵐ49ˢ on July 18, 2002.

Solution.—*Steps 1 and 2*: (2002−1972)÷4=7, remainder 2. Use column 2, and multiply quadrennial corrections by 7. Corrected values: GHA, July 16, 178°31′1+(7×0′.05)=178°31′.5; July 19, 178°27′.2+(7×0′.06)=178°27′.6. Dec., July 16, 21°27′.9N−(7×0′.41)=21°25′.0N; July 19, 20°57′.5N−(7×0′.44)=20°54′.4N.

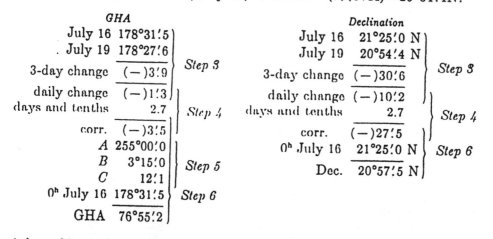

GHA			Declination		
July 16	178°31′.5		July 16	21°25′.0 N	
July 19	178°27′.6		July 19	20°54′.4 N	
3-day change	(−)3′.9	Step 3	3-day change	(−)30′.6	Step 3
daily change	(−)1′.3		daily change	(−)10′.2	
days and tenths	2.7	Step 4	days and tenths	2.7	Step 4
corr.	(−)3′.5		corr.	(−)27′.5	
A	255°00′.0		0ʰ July 16	21°25′.0 N	Step 6
B	3°15′.0	Step 5	Dec.	20°57′.5 N	
C	12′.1				
0ʰ July 16	178°31′.5	Step 6			
GHA	76°55′.2				

Aries table. 1. Subtract 1972 from the year and divide the difference by four, obtaining (*a*) a whole number, and (*b*) a remainder. Enter column indicated by remainder (*b*) and take out value for given month.

2. Enter multiplication table with whole number (*a*) of step 1, day of month, hours of GMT, minutes of GMT, and seconds of GMT, and take out corrections *D*, *E*, *F*, *G*, and *C*, respectively.

3. Add values of steps 1 and 2.

Example.—Find GHAϒ at GMT 11ʰ06ᵐ33ˢ on November 28, 1995.

Solution.—*Step* 1: (1995−1972)÷4=5, remainder 3. Use column 3.

GHAϒ		
Nov.	38°40′.6	Step 1
D	9′.2	
E	27°35′.9	
F	165°27′.1	Step 2
G	1°30′.2	
C	8′.2	
GHAϒ	233°31′.2	Step 3

Stars table. 1. Enter table with star name, and take out tabulated values.

2. Subtract 1972.0 from given year and tenth, and multiply annual correction by difference. Apply as correction (+or−, as appropriate) to value of step 1.

Example.—Find SHA and declination of Spica on September 11, 2011.

Solution.—From decimal table, September 11, 2011=2011.7. 2011.7−1972.0= 39.7.

SHA			Declination		
1972.0	159°04′.3	Step 1	1972.0	11°01′.0 S	Step 1
39.7×(−)0′.79	(−)31′.4	Step 2	39.7×0.31	(+)12′.3	Step 2
SHA	158°32′.9		Dec.	11°13′.3 S	

To determine GHA of star, add GHAϒ and SHA☆ for given time and date.

STARS

SHA (1972.0)	Annual Corr.	Star	Dec. (1972.0)	Annual Corr.
315 42.0	−0.57	Acamar	40 25.0 S	−0.24
335 49.9	−0.56	Achernar	57 22.7 S	−0.30
173 44.6	−0.84	Acrux	62 56.6 S	+0.33
291 25.3	−0.86	Aldebaran	16 27.2 N	+0.12
153 23.4	−0.59	Alkaid	49 27.2 N	−0.30
218 26.8	−0.74	Alphard	8 32.2 S	+0.26
126 37.5	−0.64	Alphecca	26 48.5 N	−0.20
358 16.0	−0.78	Alpheratz	28 56.2 N	+0.33
62 38.7	−0.73	Altair	8 47.6 N	+0.16
113 04.7	−0.92	Antares	26 22.3 S	+0.13
146 24.2	−0.68	Arcturus	19 19.6 N	−0.31
108 34.7	−1.59	Atria	68 58.7 S	+0.11
271 35.2	−0.81	Betelgeuse	7 24.2 N	+0.01
264 10.0	−0.33	Canopus	52 40.8 S	+0.03
281 20.7	−1.11	Capella	45 58.3 N	+0.06
49 52.9	−0.51	Deneb	45 10.8 N	+0.22
183 05.5	−0.76	Denebola	14 43.7 N	−0.34
349 27.2	−0.75	Diphda	18 08.4 S	−0.33
194 29.8	−0.92	Dubhe	61 54.2 N	−0.32
34 17.8	−0.74	Enif	9 44.8 N	+0.28
15 58.4	−0.83	Fomalhaut	29 46.3 S	−0.32
328 36.2	−0.85	Hamal	23 19.8 N	+0.28
137 18.4	+0.04	Kochab	74 16.2 N	−0.25
148 44.6	−0.88	Menkent	36 14.0 S	+0.29
309 25.3	−1.07	Mirfak	49 45.8 N	+0.21
76 37.1	−0.93	Nunki	26 20.0 S	−0.08
54 08.4	−1.18	Peacock	56 49.6 S	−0.19
244 05.9	−0.92	Pollux	28 05.7 N	−0.15
245 32.4	−0.78	Procyon	5 17.9 N	−0.16
96 35.5	−0.70	Rasalhague	12 34.8 N	−0.04
208 16.8	−0.80	Regulus	12 06.3 N	−0.29
281 42.1	−0.72	Rigel	8 14.0 S	−0.07
140 41.5	−1.02	Rigil Kent.	60 43.2 S	+0.25
350 16.4	−0.86	Schedar	56 23.0 N	+0.33
259 01.3	−0.66	Sirius	16 40.6 S	+0.08
159 04.3	−0.79	Spica	11 01.0 S	+0.31
223 15.5	−0.55	Suhail	43 19.1 S	+0.24
81 00.2	−0.51	Vega	38 45.4 N	+0.06

MULTIPLICATION TABLE

No.	A	B	C	D	E	F	G
1	15	0 15	0.2	1.8	0 59.1	15 02.5	0 15.0
2	30	0 30	0.5	3.7	1 58.3	30 04.9	0 30.1
3	45	0 45	0.8	5.5	2 57.4	45 07.4	0 45.1
4	60	1 00	1.0	7.4	3 56.6	60 09.9	1 00.2
5	75	1 15	1.2	9.2	4 55.7	75 12.3	1 15.2
6	90	1 30	1.5	11.0	5 54.8	90 14.8	1 30.2
7	105	1 45	1.8	12.9	6 54.0	105 17.2	1 45.3
8	120	2 00	2.0	14.7	7 53.1	120 19.7	2 00.3
9	135	2 15	2.2	16.6	8 52.3	135 22.2	2 15.4
10	150	2 30	2.5	18.4	9 51.4	150 24.6	2 30.4
11	165	2 45	2.8	20.2	10 50.5	165 27.1	2 45.5
12	180	3 00	3.0	22.1	11 49.7	180 29.6	3 00.5
13	195	3 15	3.2	23.9	12 48.8	195 32.0	3 15.5
14	210	3 30	3.5	25.8	13 48.0	210 34.5	3 30.6
15	225	3 45	3.8	27.6	14 47.1	225 37.0	3 45.6
16	240	4 00	4.0	29.4	15 46.2	240 39.4	4 00.7
17	255	4 15	4.2	31.3	16 45.4	255 41.9	4 15.7
18	270	4 30	4.5	33.1	17 44.5	270 44.4	4 30.7
19	285	4 45	4.8	35.0	18 43.7	285 46.8	4 45.8
20	300	5 00	5.0	36.8	19 42.8	300 49.3	5 00.8
21	315	5 15	5.2	38.6	20 41.9	315 51.7	5 15.9
22	330	5 30	5.5	40.5	21 41.1	330 54.2	5 30.9
23	345	5 45	5.8	42.3	22 40.2	345 56.7	5 45.9
24	360	6 00	6.0	44.2	23 39.4	360 59.1	6 01.0
25	—	6 15	6.2	46.0	24 38.5	—	6 16.0
26	—	6 30	6.5	47.8	25 37.6	—	6 31.1
27	—	6 45	6.8	49.7	26 36.8	—	6 46.1
28	—	7 00	7.0	51.5	27 35.9	—	7 01.1
29	—	7 15	7.2	53.4	28 35.1	—	7 16.2
30	—	7 30	7.5	55.2	29 34.2	—	7 31.2
31	—	7 45	7.8	57.0	30 33.3	—	7 46.3
32	—	8 00	8.0	58.9	—	—	8 01.3
33	—	8 15	8.2	60.7	—	—	8 16.4
34	—	8 30	8.5	62.6	—	—	8 31.4
35	—	8 45	8.8	64.4	—	—	8 46.4
36	—	9 00	9.0	66.2	—	—	9 01.5
37	—	9 15	9.2	68.1	—	—	9 16.5
38	—	9 30	9.5	69.9	—	—	9 31.6
39	—	9 45	9.8	71.8	—	—	9 46.6
40	—	10 00	10.0	73.6	—	—	10 01.6
41	—	10 15	10.2	75.4	—	—	10 16.7
42	—	10 30	10.5	77.3	—	—	10 31.7
43	—	10 45	10.8	79.1	—	—	10 46.8
44	—	11 00	11.0	81.0	—	—	11 01.8
45	—	11 15	11.2	82.8	—	—	11 16.8
46	—	11 30	11.5	84.6	—	—	11 31.9
47	—	11 45	11.8	86.5	—	—	11 46.9
48	—	12 00	12.0	88.3	—	—	12 02.0
49	—	12 15	12.2	90.2	—	—	12 17.0
50	—	12 30	12.5	92.0	—	—	12 32.1
51	—	12 45	12.8	93.8	—	—	12 47.1
52	—	13 00	13.0	95.7	—	—	13 02.1
53	—	13 15	13.2	97.5	—	—	13 17.2
54	—	13 30	13.5	99.4	—	—	13 32.2
55	—	13 45	13.8	—	—	—	13 47.3
56	—	14 00	14.0	—	—	—	14 02.3
57	—	14 15	14.2	—	—	—	14 17.3
58	—	14 30	14.5	—	—	—	14 32.4
59	—	14 45	14.8	—	—	—	14 47.4
60	—	15 00	15.0	—	—	—	15 02.5

ARIES (♈)

0	1	Month	2	3
98 46.2	99 31.0	Jan.	99 16.7	99 02.4
129 19.5	130 04.4	Feb.	129 50.1	129 35.7
157 54.5	157 40.3	Mar.	157 25.9	157 11.6
188 27.8	188 13.5	Apr.	187 59.2	187 44.9
218 02.0	217 47.7	May	217 33.4	217 19.0
248 35.3	248 21.0	June	248 06.7	247 52.3
278 09.5	277 55.2	July	277 40.9	277 26.5
308 42.8	308 28.5	Aug.	308 14.2	307 59.8
339 16.1	339 01.8	Sept.	338 47.5	338 33.1
8 50.3	8 36.0	Oct.	8 21.6	8 07.3
39 23.6	39 09.2	Nov.	38 54.9	38 40.6
68 57.7	68 43.4	Dec.	68 29.1	68 14.7

DECIMAL PARTS OF DAY AND YEAR

Decimal	0.0	0.1	0.2	0.3	0.4	0.5	0.6	0.7	0.8	0.9	1.0
Hour of Day	0000 to 0112	0112 to 0336	0336 to 0600	0600 to 0824	0824 to 1048	1048 to 1312	1312 to 1536	1536 to 1800	1800 to 2024	2024 to 2248	2248 to 2400
Day of Year	Jan. 1 to Jan. 18	Jan. 19 to Feb. 23	Feb. 24 to Apr. 1	Apr 2 to May 7	May 8 to June 13	June 14 to July 19	July 20 to Aug. 25	Aug. 26 to Sept. 30	Oct. 1 to Nov. 6	Nov. 7 to Dec. 12	Dec. 13 to Dec. 31

SUN

GHA −175° (0)	Dec. (0)	Quad. GHA Corr.	GHA −175° (1)	Dec. (1)	Date	GHA −175° (2)	Dec. (2)	Quad. Dec. Corr.	GHA −175° (3)	Dec. (3)
JANUARY										
4 14.4	23 05.5 S	−0.11	4 09.0	23 02.0 S	1	4 10.8	23 03.1 S	−0.32	4 12.9	23 04.2 S
3 53.3	22 50.2 S	−0.13	3 48.0	22 45.6 S	4	3 49.8	22 47.0 S	−0.35	3 51.9	22 48.4 S
3 33.0	22 30.7 S	−0.12	3 27.8	22 25.1 S	7	3 29.8	22 26.8 S	−0.39	3 31.7	22 28.6 S
3 13.8	22 07.2 S	−0.09	3 08.8	22 00.6 S	10	3 10.8	22 02.7 S	−0.42	3 12.5	22 04.8 S
2 55.7	21 39.9 S	−0.04	2 51.1	21 32.2 S	13	2 53.0	21 34.7 S	−0.44	2 54.5	21 37.1 S
2 38.9	21 08.7 S	+0.03	2 34.8	21 00.2 S	16	2 36.6	21 02.9 S	−0.46	2 37.9	21 05.6 S
2 23.7	20 33.9 S	+0.09	2 20.1	20 24.5 S	19	2 21.7	20 27.5 S	−0.48	2 22.9	20 30.5 S
2 10.1	19 55.6 S	+0.13	2 07.0	19 45.4 S	22	2 08.4	19 48.7 S	−0.49	2 09.5	19 51.9 S
1 58.2	19 13.9 S	+0.15	1 55.7	19 02.9 S	25	1 56.8	19 06.4 S	−0.52	1 57.9	19 10.0 S
1 48.2	18 29.1 S	+0.15	1 46.1	18 17.3 S	28	1 47.0	18 21.1 S	−0.54	1 48.0	18 24.9 S
FEBRUARY										
1 37.7	17 24.7 S	+0.13	1 36.1	17 11.9 S	1	1 36.8	17 16.0 S	−0.57	1 37.8	17 20.2 S
1 32.0	16 33.2 S	+0.14	1 30.7	16 19.6 S	4	1 31.3	16 24.0 S	−0.59	1 32.1	16 28.4 S
1 28.0	15 39.0 S	+0.15	1 27.1	15 24.8 S	7	1 27.8	15 29.5 S	−0.60	1 28.3	15 34.0 S
1 25.9	14 42.5 S	+0.19	1 25.4	14 27.7 S	10	1 26.0	14 32.5 S	−0.61	1 26.3	14 37.2 S
1 25.4	13 43.6 S	+0.24	1 25.4	13 28.4 S	13	1 25.9	13 33.4 S	−0.60	1 26.0	13 38.2 S
1 26.7	12 42.8 S	+0.29	1 27.2	12 27.1 S	16	1 27.5	12 32.3 S	−0.60	1 27.4	12 37.2 S
1 29.6	11 40.1 S	+0.34	1 30.6	11 24.0 S	19	1 30.7	11 29.3 S	−0.59	1 30.4	11 34.4 S
1 34.0	10 35.8 S	+0.36	1 35.6	10 19.3 S	22	1 35.4	10 24.7 S	−0.59	1 35.1	10 30.0 S
1 40.0	9 30.0 S	+0.36	1 41.9	9 13.2 S	25	1 41.5	9 18.6 S	−0.60	1 41.2	9 24.1 S
1 47.4	8 23.0 S	+0.34	1 49.5	8 05.8 S	28	1 49.0	8 11.3 S	−0.61	1 48.7	8 17.0 S
MARCH										
1 52.9	7 37.6 S	+0.32	1 52.3	7 43.1 S	1	1 51.7	7 48.7 S	−0.62	1 51.4	7 54.3 S
2 02.3	6 28.8 S	+0.31	2 01.4	6 34.4 S	4	2 00.8	6 40.0 S	−0.62	2 00.4	6 45.7 S
2 12.6	5 19.2 S	+0.31	2 11.6	5 24.8 S	7	2 11.0	5 30.5 S	−0.62	2 10.5	5 36.3 S
2 23.8	4 08.9 S	+0.33	2 22.6	4 14.5 S	10	2 22.1	4 20.4 S	−0.61	2 21.4	4 26.1 S
2 35.7	2 58.2 S	+0.36	2 34.5	3 03.8 S	13	2 33.9	3 09.7 S	−0.59	2 33.1	3 15.5 S
2 48.2	1 47.1 S	+0.40	2 47.1	1 52.8 S	16	2 46.4	1 58.7 S	−0.56	2 45.4	2 04.5 S
3 01.1	0 35.9 S	+0.43	3 00.1	0 41.7 S	19	2 59.4	0 47.6 S	−0.54	2 58.3	0 53.3 S
3 14.5	0 35.2 N	+0.44	3 13.6	0 29.4 N	22	3 12.7	0 23.6 N	+0.53	3 11.6	0 17.8 N
3 28.1	1 46.1 N	+0.43	3 27.2	1 40.3 N	25	3 26.2	1 34.6 N	+0.52	3 25.2	1 28.8 N
3 41.9	2 56.6 N	+0.40	3 40.9	2 50.9 N	28	3 39.8	2 45.2 N	+0.52	3 38.9	2 39.4 N
APRIL										
4 00.0	4 29.8 N	+0.35	3 59.0	4 24.3 N	1	3 57.9	4 18.6 N	+0.52	3 57.1	4 12.8 N
4 13.4	5 38.9 N	+0.33	4 12.3	5 33.4 N	4	4 11.2	5 27.8 N	+0.51	4 10.5	5 22.1 N
4 26.4	6 47.1 N	+0.32	4 25.2	6 41.7 N	7	4 24.3	6 36.2 N	+0.49	4 23.5	6 30.6 N
4 38.8	7 54.3 N	+0.33	4 37.7	7 49.0 N	10	4 36.9	7 43.5 N	+0.47	4 36.0	7 38.0 N
4 50.6	9 00.3 N	+0.35	4 49.6	8 55.1 N	13	4 48.9	8 49.7 N	+0.44	4 48.0	8 44.3 N
5 01.7	10 05.0 N	+0.37	5 00.9	9 59.8 N	16	5 00.2	9 54.5 N	+0.41	4 59.2	9 49.3 N
5 12.0	11 08.1 N	+0.39	5 11.3	11 03.0 N	19	5 10.6	10 57.9 N	+0.39	5 09.7	10 52.8 N
5 21.4	12 09.6 N	+0.37	5 20.9	12 04.6 N	22	5 20.0	11 59.7 N	+0.37	5 19.3	11 54.7 N
5 29.8	13 09.2 N	+0.33	5 29.4	13 04.4 N	25	5 28.5	12 59.7 N	+0.36	5 28.0	12 54.8 N
5 37.2	14 06.9 N	+0.28	5 36.8	14 02.3 N	28	5 36.0	13 57.7 N	+0.34	5 35.6	13 53.0 N
MAY										
5 43.4	15 02.6 N	+0.24	5 43.0	14 58.1 N	1	5 42.3	14 53.7 N	+0.33	5 42.1	14 49.2 N
5 48.4	15 56.0 N	+0.20	5 48.0	15 51.7 N	4	5 47.5	15 47.5 N	+0.31	5 47.3	15 43.1 N
5 52.1	16 47.0 N	+0.19	5 51.7	16 43.0 N	7	5 51.4	16 38.9 N	+0.28	5 51.3	16 34.7 N
5 54.5	17 35.5 N	+0.19	5 54.2	17 31.7 N	10	5 54.1	17 27.8 N	+0.25	5 53.9	17 23.9 N
5 55.6	18 21.4 N	+0.21	5 55.5	18 17.7 N	13	5 55.5	18 14.1 N	+0.22	5 55.3	18 10.4 N
5 55.4	19 04.5 N	+0.23	5 55.5	19 01.0 N	16	5 55.5	18 57.6 N	+0.19	5 55.4	18 54.2 N
5 53.9	19 44.7 N	+0.23	5 54.3	19 41.4 N	19	5 54.3	19 38.3 N	+0.16	5 54.3	19 35.1 N
5 51.3	20 21.8 N	+0.20	5 51.8	20 18.8 N	22	5 51.7	20 15.9 N	+0.13	5 51.9	20 13.0 N
5 47.5	20 55.8 N	+0.16	5 48.1	20 53.1 N	25	5 48.0	20 50.5 N	+0.11	5 48.4	20 47.7 N
5 42.6	21 26.6 N	+0.11	5 43.2	21 24.2 N	28	5 43.2	21 21.8 N	+0.08	5 43.8	21 19.3 N
JUNE										
5 34.5	22 02.4 N	+0.04	5 35.1	22 00.4 N	1	5 35.2	21 58.3 N	+0.04	5 36.0	21 56.2 N
5 27.4	22 25.3 N	+0.02	5 27.9	22 23.5 N	4	5 28.2	22 21.7 N	+0.01	5 28.9	22 19.9 N
5 19.4	22 44.6 N	+0.02	5 19.9	22 43.1 N	7	5 20.4	22 41.6 N	−0.02	5 21.1	22 40.1 N
5 10.7	23 00.3 N	+0.04	5 11.3	22 59.1 N	10	5 11.9	22 57.9 N	−0.05	5 12.5	22 56.7 N
5 01.5	23 12.4 N	+0.07	5 02.3	23 11.5 N	13	5 02.9	23 10.6 N	−0.08	5 03.4	23 09.7 N
4 51.9	23 20.8 N	+0.09	4 52.8	23 20.2 N	16	4 53.4	23 19.6 N	−0.12	4 53.9	23 19.0 N
4 42.1	23 25.5 N	+0.08	4 43.2	23 25.3 N	19	4 43.6	23 24.9 N	−0.15	4 44.2	23 24.6 N
4 32.4	23 26.5 N	+0.06	4 33.4	23 26.6 N	22	4 33.7	23 26.5 N	−0.18	4 34.5	23 26.5 N
4 22.8	23 23.8 N	+0.01	4 23.8	23 24.1 N	25	4 24.0	23 24.4 N	−0.22	4 24.8	23 24.6 N
4 13.4	23 17.4 N	−0.03	4 14.3	23 18.0 N	28	4 14.5	23 18.5 N	−0.25	4 15.4	23 19.1 N

SUN										
0		Quad. GHA Corr.	**1**		Date	**2**		Quad. Dec. Corr.	**3**	
GHA −175°	Dec.		GHA −175°	Dec.		GHA −175°	Dec.		GHA −175°	Dec.
JULY										
4 04.5	23 07.3 N	−0.06	4 05.2	23 08.2 N	1	4 05.4	23 09.0 N	−0.28	4 06.3	23 09.9 N
3 56.1	22 53.5 N	−0.07	3 56.6	22 54.7 N	4	3 57.0	22 55.8 N	−0.31	3 57.7	22 57.0 N
3 48.4	22 36.2 N	−0.05	3 48.8	22 37.6 N	7	3 49.2	22 39.1 N	−0.34	3 49.8	22 40.5 N
3 41.5	22 15.3 N	−0.01	3 41.9	22 17.1 N	10	3 42.3	22 18.8 N	−0.37	3 42.7	22 20.5 N
3 35.5	21 51.0 N	+0.03	3 36.0	21 53.0 N	13	3 36.2	21 55.0 N	−0.39	3 36.4	21 57.0 N
3 30.6	21 23.3 N	+0.05	3 31.1	21 25.7 N	16	3 31.1	21 27.9 N	−0.41	3 31.3	21 30.1 N
3 26.9	20 52.4 N	+0.06	3 27.4	20 55.0 N	19	3 27.2	20 57.5 N	−0.44	3 27.4	21 00.0 N
3 24.5	20 18.3 N	+0.04	3 24.9	20 21.2 N	22	3 24.5	20 23.8 N	−0.47	3 24.7	20 26.6 N
3 23.4	19 41.1 N	+0.01	3 23.6	19 44.2 N	25	3 23.0	19 47.1 N	−0.50	3 23.3	19 50.2 N
3 23.6	19 01.0 N	−0.02	3 23.6	19 04.3 N	28	3 23.0	19 07.5 N	−0.54	3 23.1	19 10.8 N
AUGUST										
3 26.0	18 03.2 N	−0.03	3 25.6	18 06.8 N	1	3 25.1	18 10.3 N	−0.57	3 25.1	18 13.9 N
3 29.3	17 16.7 N	−0.02	3 28.8	17 20.5 N	4	3 28.2	17 24.2 N	−0.59	3 28.0	17 28.0 N
3 33.9	16 27.7 N	+0.02	3 83.4	16 31.6 N	7	3 32.8	16 35.6 N	−0.60	3 32.3	16 39.6 N
3 39.9	15 36.3 N	+0.06	3 39.3	15 40.4 N	10	3 38.6	15 44.6 N	−0.60	3 38.0	15 48.7 N
3 47.1	14 42.6 N	+0.09	3 46.5	14 47.0 N	13	3 45.7	14 51.3 N	−0.61	3 44.9	14 55.6 N
3 55.7	13 46.8 N	+0.11	3 55.0	13 51.4 N	16	3 54.0	13 55.8 N	−0.62	3 53.2	14 00.3 N
4 05.4	12 49.1 N	+0.11	4 04.7	12 53.9 N	19	4 03.4	12 58.4 N	−0.64	4 02.7	13 03.1 N
4 16.2	11 49.6 N	+0.09	4 15.4	11 54.5 N	22	4 14.0	11 59.1 N	−0.66	4 13.3	12 04.0 N
4 28.1	10 48.4 N	+0.06	4 27.2	10 53.3 N	25	4 25.7	10 58.1 N	−0.68	4 24.9	11 03.2 N
4 40.9	9 45.6 N	+0.03	4 39.8	9 50.6 N	28	4 38.3	9 55.6 N	−0.70	4 37.5	10 00.8 N
SEPTEMBER										
4 59.2	8 19.8 N	+0.02	4 57.9	8 24.9 N	1	4 56.5	8 30.0 N	−0.71	4 55.6	8 35.4 N
5 13.7	7 13.9 N	+0.04	5 12.3	7 19.2 N	4	5 10.9	7 24.4 N	−0.70	5 09.9	7 29.8 N
5 28.7	6 07.0 N	+0.08	5 27.3	6 12.4 N	7	5 26.0	6 17.8 N	−0.69	5 24.7	6 23.2 N
5 44.1	4 59.2 N	+0.11	5 42.8	5 04.7 N	10	5 41.4	5 10.1 N	−0.68	5 40.1	5 15.6 N
5 59.9	3 50.7 N	+0.13	5 58.6	3 56.2 N	13	5 57.1	4 01.7 N	−0.67	5 55.8	4 07.2 N
6 15.8	2 41.5 N	+0.13	6 14.7	2 47.1 N	16	6 13.0	2 52.6 N	−0.67	6 11.7	2 58.2 N
6 31.9	1 31.9 N	+0.11	6 30.7	1 37.5 N	19	6 29.0	1 43.0 N	−0.68	6 27.8	1 48.7 N
6 47.9	0 21.9 N	+0.08	6 46.6	0 27.6 N	22	6 44.9	0 33.1 N	−0.68	6 43.8	0 38.8 N
7 03.6	0 48.1 S	+0.04	7 02.3	0 42.5 S	25	7 00.6	0 37.0 S	+0.69	6 59.6	0 31.2 S
7 19.0	1 58.3 S	+0.01	7 17.6	1 52.7 S	28	7 16.1	1 47.1 S	+0.68	7 15.1	1 41.3 S
OCTOBER										
7 33.8	3 08.2 S	0.00	7 32.4	3 02.7 S	1	7 31.0	2 57.1 S	+0.67	7 30.0	2 51.4 S
7 48.0	4 17.9 S	+0.02	7 46.6	4 12.4 S	4	7 45.4	4 06.8 S	+0.65	7 44.3	4 01.2 S
8 01.3	5 27.2 S	+0.05	8 00.1	5 21.7 S	7	7 59.0	5 16.1 S	+0.62	7 57.9	5 10.5 S
8 13.7	6 35.8 S	+0.07	8 12.7	6 30.3 S	10	8 11.6	6 24.8 S	+0.60	8 10.5	6 19.3 S
8 25.1	7 43.6 S	+0.08	8 24.3	7 38.1 S	13	8 23.2	7 32.8 S	+0.57	8 22.2	7 27.3 S
8 35.4	8 50.5 S	+0.07	8 34.7	8 45.0 S	16	8 33.6	8 39.8 S	+0.56	8 32.8	8 34.4 S
8 44.4	9 56.2 S	+0.04	8 43.8	9 50.9 S	19	8 42.8	9 45.7 S	+0.54	8 42.2	9 40.3 S
8 52.0	11 00.6 S	−0.01	8 51.5	10 55.4 S	22	8 50.5	10 50.3 S	+0.53	8 50.1	10 45.0 S
8 58.1	12 03.5 S	−0.05	8 57.6	11 58.5 S	25	8 56.9	11 53.5 S	+0.52	8 56.6	11 48.3 S
9 02.6	13 04.7 S	−0.08	9 02.1	12 59.9 S	28	9 01.6	12 55.0 S	+0.50	9 01.5	12 50.0 S
NOVEMBER										
9 05.8	14 23.5 S	−0.09	9 05.5	14 18.9 S	1	9 05.4	14 14.2 S	+0.46	9 05.3	14 09.4 S
9 06.1	15 20.3 S	−0.07	9 06.0	15 15.8 S	4	9 06.1	15 11.2 S	+0.42	9 06.0	15 06.6 S
9 04.6	16 14.7 S	−0.04	9 04.7	16 10.4 S	7	9 05.0	16 06.0 S	+0.38	9 04.9	16 01.6 S
9 01.1	17 06.7 S	−0.03	9 01.6	17 02.5 S	10	9 01.9	16 58.4 S	+0.35	9 02.0	16 54.2 S
8 55.8	17 56.1 S	−0.03	8 56.5	17 52.1 S	13	8 56.9	17 48.2 S	+0.31	8 57.2	17 44.2 S
8 48.6	18 42.6 S	−0.05	8 49.5	18 38.8 S	16	8 49.9	18 35.2 S	+0.28	8 50.5	18 31.4 S
8 39.6	19 26.1 S	−0.09	8 40.6	19 22.7 S	19	8 41.1	19 19.2 S	+0.25	8 41.9	19 15.7 S
8 28.7	20 06.5 S	−0.14	8 29.8	20 03.3 S	22	8 30.5	20 00.1 S	+0.22	8 31.5	19 56.9 S
8 16.1	20 43.6 S	−0.18	8 17.2	20 40.7 S	25	8 18.1	20 37.8 S	+0.19	8 19.3	20 34.8 S
8 01.7	21 17.2 S	−0.20	8 02.8	21 14.6 S	28	8 04.1	21 11.9 S	+0.16	8 05.3	21 09.3 S
DECEMBER										
7 45.6	21 47.3 S	−0.18	7 46.9	21 44.9 S	1	7 48.4	21 42.5 S	+0.12	7 49.6	21 40.1 S
7 28.1	22 13.5 S	−0.15	7 29.6	22 11.5 S	4	7 31.3	22 09.4 S	+0.08	7 32.5	22 07.3 S
7 09.3	22 35.9 S	−0.11	7 11.0	22 34.2 S	7	7 12.8	22 32.4 S	+0.04	7 13.9	22 30.6 S
6 49.3	22 54.3 S	−0.09	6 51.3	22 52.9 S	10	6 53.0	22 51.4 S	0.00	6 54.3	22 50.0 S
6 28.5	23 08.6 S	−0.08	6 30.6	23 07.6 S	13	6 32.3	23 06.4 S	−0.04	6 33.7	23 05.3 S
6 06.9	23 18.8 S	−0.09	6 09.1	23 18.1 S	16	6 10.8	23 17.3 S	−0.08	6 12.4	23 16.5 S
5 44.9	23 24.8 S	−0.12	5 47.1	23 24.4 S	19	5 48.8	23 24.0 S	−0.12	5 50.5	23 23.6 S
5 22.6	23 26.6 S	−0.15	5 24.7	23 26.6 S	22	5 26.5	23 26.5 S	−0.16	5 28.3	23 26.4 S
5 00.2	23 24.1 S	−0.17	5 02.2	23 24.4 S	25	5 04.1	23 24.7 S	−0.19	5 05.9	23 25.0 S
4 38.0	23 17.5 S	−0.17	4 39.9	23 18.1 S	28	4 41.9	23 18.7 S	−0.23	4 43.6	23 19.3 S

Appendix D.

A Simplified Mercator Chart Construction.

1. Select a convenient length to represent 1° of longitude.
   We suggest 12 cms or 6 inches to represent 1° longitude.

2. Then 1' (one minute) of longitude measures:

   12 cm divided by 60 = 0.2 cm. = 1' of longitude.

   - If you prefer inches, then we recommend the clearrplastic
   (blue) rulers made by Helix, which measure half the length
   in 16ths and the other half in tenths of an inch. These are
   obtainable at Office-Max stores.   In this case a scale of
   6 inches divided by 60 gives 0.1" = 1' longtiude.

3. To calculate the corresponding measurements on the latitude
   scale, determine the length of 1° of latitude in the desired
   region by dividing the length of 1° of longitude, e.g. 12 cms
   or 6 inches, by the cosine of the latitude desired.

Example:

For latitude 37° N.

1° of lat. = 12 cms divided by cosine 37°

= 15.02563 cms.

Therefore 1' = 15.02563 divided by 60 = 0.25043 cm

= 1 nautical mile.

THIS FORMULA IS DERIVED AS FOLLOWS:

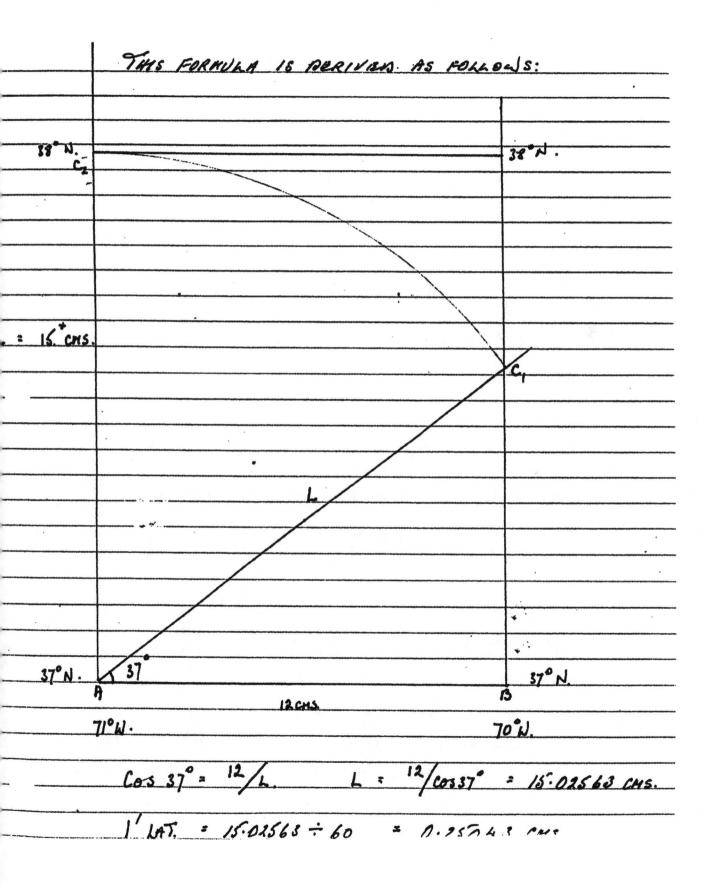

$$\text{Cos } 37° = ^{12}/_{L.} \qquad L = ^{12}/_{\text{Cos } 37°} = 15.02563 \text{ CMS.}$$

$$1' \text{ LAT.} = 15.02563 \div 60 = 0.25043 \text{ CMS.}$$

What are your Coordinates if your position bears 055° T and 17 nautical miles from a datum of 40° N., 75° W ?

Use a 12 cm. scale.

Therefore 1' of longitude = 0.2 cms.

1' of latitude = 0.26108 cms

17 n.m. = 17 x 0.26108 = 4.44 cm.

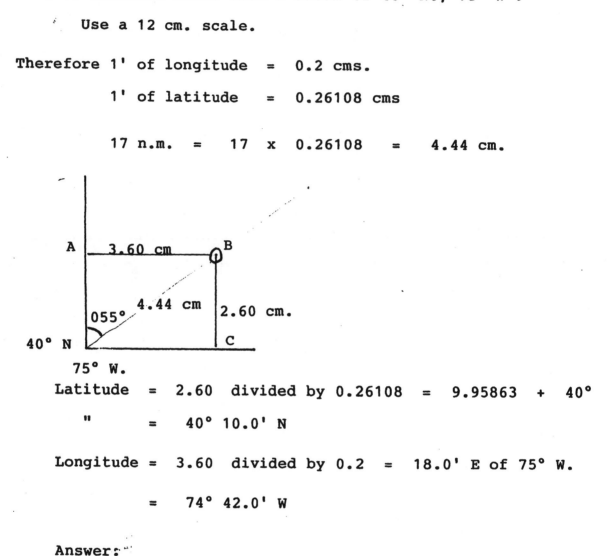

Latitude = 2.60 divided by 0.26108 = 9.95863 + 40°

" = 40° 10.0' N

Longitude = 3.60 divided by 0.2 = 18.0' E of 75° W.

= 74° 42.0' W

Answer:

Your position = 40° 10.0' N; 74° 42.0' W

Note:

The coordinate components AB and BC may be derived by measurement with a centimeter ruler, or calculated by simple trig., since you have the angle and the hypotenuse.

What is the bearing and the distance of your position from a datum of 35° N., 76° W., if you are located at 35° 35.0' N 75° 15.0' W ?

1' longt. @ 35.5° N   =   0.2 cms.

1' lat.   @ 35.5° N   =   0.24567 cms.

35.0' lat   x   0.24567   =   8.59829 cms N.

15' longt.   x   0.2   =   3.0 cms W of 75° W.

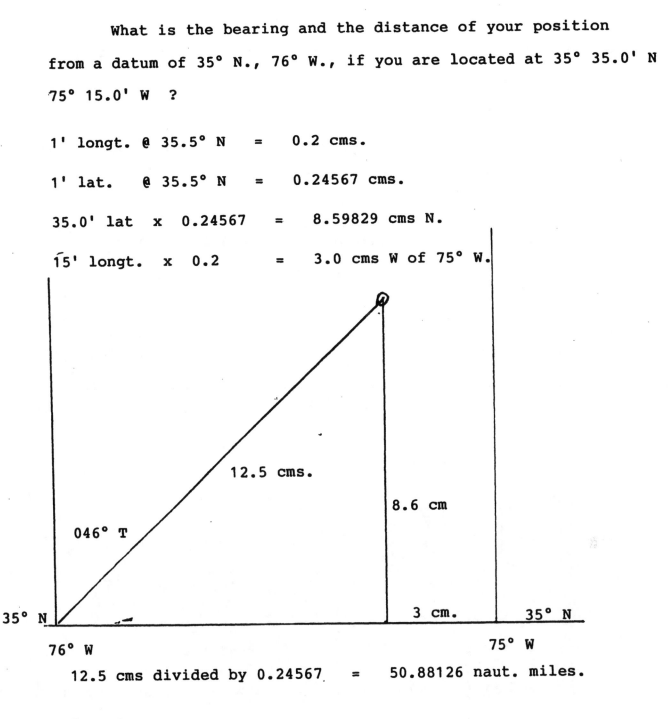

12.5 cms.

8.6 cm

046° T

35° N   3 cm.   35° N

76° W   75° W

12.5 cms divided by 0.24567   =   50.88126 naut. miles.

Answer:
          046° T /   50.9 n.m.

You are sailing on a course of 123° T at 8 knots. At 0800 GMT you obtain a star sight which calculates out to have an azimuth of 035° T, with an intercept of 10' towards, from an assumed position of 25° N., 90° W.

At 0845 GMT another star sight, using the same assumed position, gives an azimuth of 297° T, and an intercept of 9' away.

Determine your running fix at 0845 GMT, using a Mercator projection with a scale of 12 cm representing 1° of longitude.

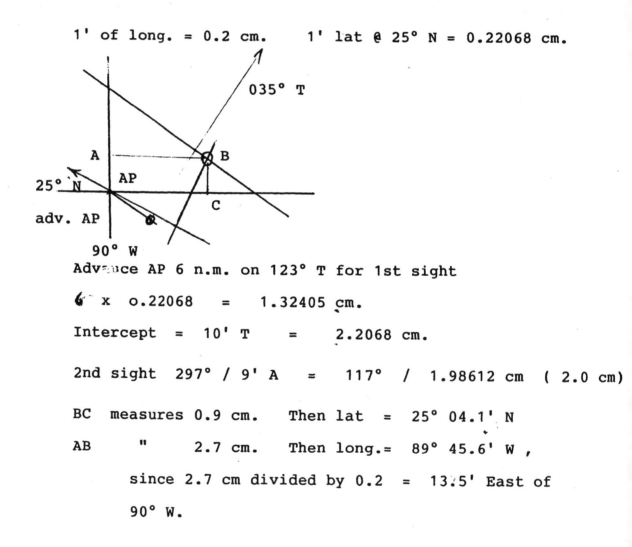

1' of long. = 0.2 cm.   1' lat @ 25° N = 0.22068 cm.

035° T

A   B

25° N   AP

adv. AP

C

90° W

Advance AP 6 n.m. on 123° T for 1st sight

6 x 0.22068   =   1.32405 cm.

Intercept = 10' T   =   2.2068 cm.

2nd sight 297° / 9' A   =   117° / 1.98612 cm (2.0 cm)

BC measures 0.9 cm.   Then lat = 25° 04.1' N

AB   "   2.7 cm.   Then long.= 89° 45.6' W ,

    since 2.7 cm divided by 0.2 = 13.5' East of
    90° W.

Answer:

    0845 GMT running fix = 25° 04.1' N, 89° 45.6' W.

About the Author

Dennis A.J. Morey was born in Eltham, London, England in 1920, and attended private boys school there. He matriculated at the University of London in early 1939, and enlisted in the Royal Air Force in Sept. 1939. After basic training at Hastings, he was sent to the R.A.F. Navigation school in Prestwick, Scotland. Following this he attended air gunnery school and bombardier school. A short course at the Flying Boat school in Stranraer, Scotland, and then transferred to a Beaufort Torpedo-Bomber Squadron in East Anglia, where he carried out 37 missions from the Arctic Ocean to the South of France. He was then commissioned and sent to Canada for a Specialist Navigation course. On completion he was issued a First Class Navigators Certificate, and elected to the Royal Meteorological Society. He spent 2 years teaching there, and then returned to England, taking courses at the Bomber Command Conversion School, Pathfinder School and further training in airborne radar navigation and bombing. He was then sent to a Lancaster Heavy Bomber Squadron and qualified as a Pathfinder with promotion to Squadron Leader. He completed 27 missions over Germany marking the targets for the thousand bomber raids, and was awarded two Distinguished Flying crosses, He was then sent to Gander, Newfoundland, as Senior Navigation Officer and Executive Officer. Discharged in Jan. 1946 he attended medical school at the University of Western Ontario, Canada, and graduated in 1950, having been elected to the Alpha Omega Alpha Honor Medical Society and the Hippocratic Honor Society. He then took three years of training in the specialty of Internal Medicine, and a fourth year in the specialty of Gastroenterology. Following board certification in these two specialties he was elected to the Fellowship in the American College of Physicians, and appointed as Assistant Professor of Clinical Medicine at the Medical college of Virginia.

In the mid 1960's he joined the Richmond Power Squadron, a unit of the United States Power Squadron, taking all their courses and receiving the rating of Senior Navigator. He taught advanced navigation and meteorology for several years, and designed a number of navigational contests. During this time he wrote a text entitled "The Magnetic Compass, and its correction and Maintenance".

Currently he is on his second career as an historian. He is a board member of the Henricus Foundation, and chairman of the historical research committee. He has completed 13 monographs on early American history, which have been published by the Henricus Foundation, and are on sale in their gift shop. At this time he is preparing a paper entitled "Who introduced Democracy to the New World', which he is to present to the meeting of the Descendants of the Ancient Planters, in Williamsburg in late October.

A copy of the "About the Author" in the above monographs is below

About The Author

Dennis A. J. Morey, M.D. was born in London, England in 1920 and came to maturity just as the Third Reich was beginning an all-out assault upon his homeland. He matriculated to the University of London in 1939 and then enlisted in the Royal Air Force. Trained as a navigator, bomb-aimer and gunner, he was attached to #22 torpedo bomber squadron, Coastal Command, where he flew 37 missions.

After posting to Canada to teach a Specialist Navigation course, he returned to England in 1944 and was posted to #7 Lancaster Pathfinder Squadron, Bomber Command, where he flew 27 missions over occupied Europe and Germany. His wartime service against the Third Reich earned him two Distinguished Flying Crosses and the rank of Squadron Leader.

After World War II, Morey attended medical school at the University of Western Ontario, Canada. He graduated with his medical degree in 1950, and then served in a variety of posts in Florida, Texas and Richmond, Virginia, specializing in gastroenterology and internal medicine. From 1956 through 1989, he practiced medicine in Richmond and served as chief of staff for Henrico Doctors' Hospital in 1984. Dr. Morey was also elected to the Alpha Omega Alpha and Hippocratic Honor Medical societies.

In addition to his practice and administrative duties, Dr. Morey has been a member of the Richmond Academy of Medicine's Ethics Committee, and is chairman of the Ethics Committee for Henrico Doctors' Hospital. Dr. Morey also served as assistant professor of clinical medicine at the Medical College of Virginia Hospitals.

He shares his wartime education by teaching meteorology, navigation and compass correction for the Richmond Power Squadron, and mathematics and electronics for the Richmond Amateur Radio Club. A former college letterman in fencing, he has served as a fencing coach for Virginia Commonwealth University, and was a founding member of the Richmond Rugby Football Club.

As a member of the Henricus Foundation Board of Trustees, and as chairman of Historical Research, Dr. Morey has been extremely active as a speaker for the foundation. His meticulous research earned him the prestigious History Award Medal in 1996 from the National Society of the Daughters of the American Revolution.
Dr. Morey has generously donated the proceeds of his copyrighted publications to the Henricus Foundation.

Theodore W. Throckmorton
RESUME
(1)   BS degree from Richmond Professional Institute, extension of William & Mary College. Attended William & Mary College in Williamsburg one year.

(2)   Taught night school two years in Mathematics of Accounting.

(3)   Served in U.S Navy Reserve in World War II.

(4)   Opened Virginia Yacht Sales, Inc. in 1950 and served as President until 1995.

(5)   Member of Richmond Sail & Power Squadron for fifty years and also served as Commander.
Rating: Senior Navigator.

(6)   Member of Coast Guard Auxiliary.

(7)   In 1957 became Instrument and Multi Engine rated as a pilot using this to cover the East Coast of U.S. to Florida.

(8)   Received a Captains license to operate in the Atlantic area.